智能光电技术应用专业英语

主　编　王凌波　吴天慧
副主编　周　利　王九程　王　琛
参　编　张　蕾　颜红梅　宋露露
　　　　敖开发　刘晓竹
主　审　吴晓红

北京理工大学出版社
BEIJING INSTITUTE OF TECHNOLOGY PRESS

内 容 简 介

　　本书是以光电技术行业对复合人才的需求为导向编写的，内容涉及电学、光学、激光基础、光电探测技术、激光加工技术，以及光纤通信这六个方面的内容。全书共分为 6 个单元，每个单元包括 6 个模块，即技术原理和设备认知、技术应用理解、技能磨砺、任务挑战、输出与评价，以及知识拓展。

　　本书结构合理，取材广泛，文图并茂，可作为高等职业院校、高等专科院校、成人高校、民办高校、本科院校的二级职业技术学院和中等职业院校光电技术应用专业的教学用书，也可以作为相关工程技术人员和社会从业人员的参考用书和培训用书。

图书在版编目（CIP）数据

智能光电技术应用专业英语 / 王凌波，吴天慧主编.
-- 北京：北京理工大学出版社，2024.1
ISBN 978 - 7 - 5763 - 3453 - 1

Ⅰ. ①智… Ⅱ. ①王… ②吴… Ⅲ. ①光电技术 - 英语 Ⅳ. ①TN2

中国国家版本馆 CIP 数据核字（2024）第 004512 号

责任编辑：多海鹏　　　　文案编辑：辛丽莉
责任校对：周瑞红　　　　责任印制：施胜娟

出版发行 / 北京理工大学出版社有限责任公司
社　　址 / 北京市丰台区四合庄路 6 号
邮　　编 / 100070
电　　话 / (010) 68914026（教材售后服务热线）
　　　　　　　(010) 68944437（课件资源服务热线）
网　　址 / http：//www.bitpress.com.cn

版 印 次 / 2024 年 1 月第 1 版第 1 次印刷
印　　刷 / 北京广达印刷有限公司
开　　本 / 787 mm × 1092 mm　1/16
印　　张 / 14.75
彩　　插 / 2
字　　数 / 360 千字
定　　价 / 68.00 元

前　言

随着光电技术的迅速发展和广泛应用，对光电技术领域的人才需求也不断增加。为满足人才培养的需求，实现对学习者的专业、语言和信息素养的三维培养，自 2011 年起，我们成功开发了"光电子技术专业英语"教材，该教材在广大师生中获得了广泛认可。随着教育信息化建设的推进和职业教育改革的不断深入，我们于 2018 年对该教材进行了改版。如今，我们将其升级为融媒体教材"智能光电技术应用专业英语"，新版教材配套了微课、动画、思维导图、PPT 等丰富的信息化资源。通过系统地学习光电技术领域专业术语、理论知识和实践技能，学习者将能够高效地与国际同行合作、阅读科技文献、参与学术交流，并提高其在光电技术领域的职业竞争力。

本教材涉及电学、光学、激光基础、光电探测技术、激光加工技术和光纤通信这六个方面的专业内容。全书共六个单元，每个单元包括六个模块。

1. 技术原理和设备认知（Technical Principles and Device Cognition）：主要介绍光电领域的技术原理和相关器件与设备，以加强学习者对专业知识的理解和掌握。

2. 技术应用理解（Technology Application Understanding）：介绍光电技术在实际工程中的应用场景和具体用途，旨在提高学习者的解决实际问题的能力。

3. 技能磨砺（Skill Honing）：介绍有效的翻译方法和技巧，以提高翻译质量和效率。帮助学习者正确理解和转换不同领域的专业术语，确保翻译得准确和专业。通过掌握翻译技巧，学习者可以在实际任务中独立思考并采取适当的解决方法，提高解决问题的能力和学习的自信心。

4. 任务挑战（Job Task Challenges）：包括职场口语和写作任务。口语任务帮助学习者练习口语表达和沟通技巧，使他们能够流利地用英语进行日常对话、讨论和演示等，提高在不同情境下灵活运用语言进行表达和交流的能力。写作任务帮助学习者准确地表达自己的观点、想法和意见，提高书面表达能力，有助于培养逻辑思维和分析能力。

5. 输出与评价（Output and Evaluation）：引导学习者根据本单元所学知识，表达相关领域的专业信息和观点，并通过思维导图和微视频来展示。这个模块配备了任务评价表。

6. 知识拓展（Knowledge Expansion）：准备了技术文献供学习者参考，以拓展学习者对相关领域技术发展和应用的认识和理解。通过研读技术文献，学习者可以进一步提升专业能力，拓宽视野。

学习者可以通过智慧职教平台加入教材配套云课程，授课教师可以引用相关资源制作教师自己的 SPOC 课程。

本教材 Technical Principles and Device Cognition，Technology Application Understanding，Knowledge Expansion 模块的分工：Unit1 由武汉职业技术学院王凌波编写，Unit2 由王凌波、

武汉大学周利编写，Unit3 由武汉软件工程职业学院王九程编写，Unit4 由王凌波、汉口学院王琛、湖北大学张蕾编写，Unit5 由武汉软件工程职业学院颜红梅、吴天慧编写，Unit6 由王琛、武汉市仪表电子学校刘晓竹、武汉职业技术学院宋露露编写。Skill Honing 和 Job Task Challenges 模块由吴天慧编写；Output and Evaluation 模块由王凌波、张蕾、王琛开发与编写；武汉光驰教育科技股份有限公司的敖开发在教材的编写过程中，关于光电实训设备的原理与操作，提供了丰富的文档资源与专业的技术支持。本教材由王凌波和吴天慧共同完成统稿和编排；主审是武汉职业技术学院吴晓红教授。

由于编者水平有限，疏漏和不妥之处在所难免，恳请读者不吝指正。

编　者

Contents

Unit 1　Basic Circuit and Its Applications

In this unit, you are going to learn the following contents:

Task Ⅰ　Technical Principles and Device Cognition

1. 1　Basic Circuit Concepts

1. 2　Capacitance and Inductance

Task Ⅱ　Technology Application Understanding

1. 3　Voltage Dividers

1. 4　Filters

Task Ⅲ　Skill Honing

1. 5　Features of Scientific English

Task Ⅳ　Job Task Challenges

1. 6　Career Speaking: Job Interview

1. 7　Writing: Resume

Task Ⅴ　Output and Evaluation

Task Ⅵ　Knowledge Expansion

1. 8　Ohm's Law

1. 9　What Is Electricity?

Unit 1

Basic Circuit and Its Applications

Task I Technical Principles and Device Cognition

1.1 Basic Circuit Concepts

Learning Objectives

In this section, you will:

1. understand the meaning of direct current (DC).

2. describe the block diagram Fig. 1. 2.

3. learn about voltage, current, and resistance.

4. master the relationship between voltage, current, and resistance in an electrical circuit.

Warm-up Activity

Do you know the names of the components in Fig. 1. 1?

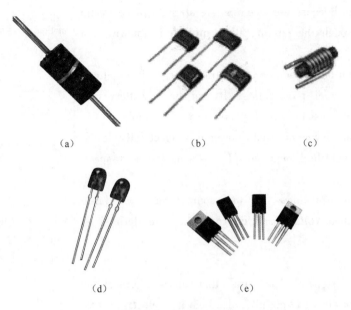

（a）　　　　　　　（b）　　　　　　　（c）

（d）　　　　　　　（e）

Fig. 1. 1　Components

 Text

Figure 1. 2 shows the basic type of electrical **circuit**, in the form of a **block diagram**. It consists of a source of electrical energy, some sort of **load** to make use of that energy, and electrical conductors connecting the source and the load.

circuit *n.* 电路
block diagram 结构图，方块图，简图
load *n.* 负载

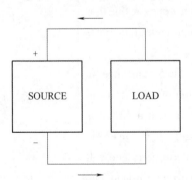

Fig. 1. 2　The basic type of electrical circuit

The electrical source has two **terminals, designated** positive (+) and negative (−). **As long as** there is an unbroken connection from source to load and back again as shown in Fig. 1. 2, **electrons** will be pushed from the negative terminal of the source, through the load, and then back to the positive terminal of the source. The arrows show the direction of electron current flow

terminal *n.* 终端，接线端
designate *v.* 指定，指派
as long as 只要，在……的时候
electron *n.* 电子

through this circuit. Because the electrons are always moving in the same direction through the circuit, their **motion** is known as a direct current.

motion *n.* 运动，动作

The source can be any source of electrical energy. In practice, there are three general possibilities. It can be a battery, an electrical generator, or some sort of electronic power supply.

The load is any device or circuit powered by **electricity**. It can be as simple as a light bulb or as complex as a modern high-speed computer.

electricity *n.* 电流，电，电学

The electricity provided by the source has two basic characteristics, called **voltage** and **current**. They are **defined** as follows.

voltage *n.* ［电工］电压
current *n.* 电流
define *vt.* 定义

Voltage

Voltage is the electrical "pressure" that causes free electrons to travel through an electrical circuit, also known as **electromotive force** (**emf**). It is measured in volts (V).

electromotive force 电动势

Current

The amount of electrical charge (the number of free electrons) moving past a given point in an electrical circuit per unit of time. Current is measured in amperes (A).

Resistance

The load, in turn, has a characteristic called resistance. By definition, that characteristic of a medium which opposes the flow of electrical current through itself. Resistance is measured in ohms (Ω).

The relationship between voltage, current, and resistance in an electrical circuit is fundamental to the operation of any circuit or device. **Verbally**, the amount of current flowing through a circuit is directly proportional to the applied voltage and inversely proportional to the circuit resistance. By explicit definition, one volt of electrical pressure can push one ampere of current through one ohm of resistance. Two volts can either push one ampere through a resistance of two ohms, or can push two amperes through one ohm. Mathematically,

verbally *adv.* 用言辞地，口头地

$$E = I \times R$$

where E is the applied voltage, or EMF;

I is the circuit current;

R is the resistance in the circuit.

Get to Work

Ⅰ. Matching

Directions: Match the following terms to appropriate definition or expression.

1. current	A. electromotive force
2. ampere	B. the flow of electrons
3. voltage	C. the unit of resistance
4. ohm	D. the unit in which current is measured
5. DC	E. the direction of current is constant

Ⅱ. Understanding Checking

Directions: Mark Y (for YES) if the statement agrees with the information given in the text; N (for NO) if the statement contradicts the information given in the text.

(   ) 1. One kilovolt equals one thousand volts.

(   ) 2. The flow of electrons through a conductor is called resistance.

(   ) 3. Current flow is represented by the letter symbol *I*.

(   ) 4. The opposition to current is called electrical resistance.

(   ) 5. The relationship between voltage, current, and resistance in an electrical circuit is fundamental to the operation of any circuit or device.

Ⅲ. Passage Completion

Directions: Fill in each of the blanks with one of the words or expressions in the box, making changes if necessary.

constant	cycle	change	negative	repeat

To have a better understanding of alternating current and voltage, it is desirable to begin with a consideration of the general situation. A function of time is called alternating when after a time interval or period, it repeats a previous succession of positive and 1. _____ valves. In other words, such an action 2. _____ itself exactly over equal intervals of time. So an alternating current is a rate of electricity which does not have a 3. _____ value in time but grows to a maximum value, decreases, 4. _____ its direction reaches a maximum value in the new direction, returns to its original value and then repeats this 5. _____ an indefinite number of times.

Ⅳ. Translation

Directions: Complete the sentences by translating into English according to the Chinese given in brackets.

1. Modern advances in the fields of computer, control system communication _____ (与电子工业密切相关).

2. When it is desirable to express a magnitude of current _____ (比安培小), the milliampere and microampere units are used.

3. _____ （电流、电压和电阻的关系） in an electrical circuit is fundamental to the operation of any circuit or device.

4. There are _____ （大量的符号） which represent an equally large range of electronic components.

5. The _____ （电流的单位是） is ampere.

1.2 Capacitance and Inductance

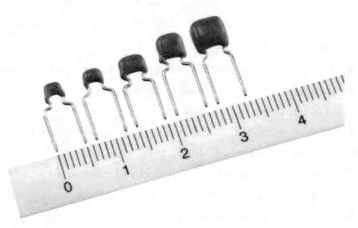

Learning Objectives

In this section, you will:

1. understand the charging process of a capacitor.

2. learn about three important factors of a capacitor.

3. comprehend the use of capacitor in a radio receiver.

4. learn how to make an inductor by hand.

5. calculate how much energy an inductor can store.

6. learn about the units of inductance.

Warm-up Activity

Do you know the names of the components in Fig. 1.3?

（a） （b）

Fig. 1.3 Components

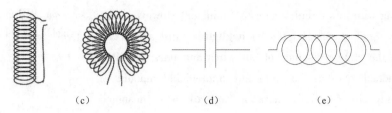

（c）　　　　　（d）　　　　（e）

Fig. 1.3　Components（Continued）

 Text

Capacitance

Electrical energy can be stored in an **electric** field. The **device** to be capable of doing this is called a **capacitor** or a condenser.

A simple capacitor consists of two metallic plates separated by a **dielectric**. If a capacitor is connected to a battery, the electrons will flow out of the negative terminal of the battery and accumulate on the capacitor plate connected to that side. At the same time, the electrons will leave the plate connected to the positive terminal and flow into the battery to make the potential difference just the same as that of the battery. Thus, the capacitor is said to be **charged**.

The capacitance is directly **proportional** to the dielectric constant of the material and to the area of the plates and inversely to the distance of the plates. It is measured in **farads**. When a change of one volt per second across it causes the current of one ampere to flow, the capacitor is said to have the capacitance of one farad. However, farad is too large a unit to be used in radio calculation, so microfarad (one millionth of a farad) and the picofarad (10^{-12} farad) are generally used.

The amount of the stored energy of a charged capacitor is proportional to the applied voltage and its capacitance. The capacitance of a capacitor is determined by three important factors, namely the area of the plate surface, the space between them and dielectric material. The larger the plate area, the smaller the space between them, the greater the capacitance.

One of the capacitors to be used in radio receiver is a variable capacitor, whose capacitance can be varied by turning the plates. It is used in the receiver for turning and varying capacitance in the circuit so as to pick up the desired signals of different wavelengths.

Inductance

Any current will create a **magnetic** field, so in fact every

electrical *adj.* 电的，有关电的
electric *adj.* 电动的，电气，以电为动力的
device *n.* 装置，设备
capacitor *n.* 电容器
dielectric *n.* 电介质，绝缘体

charge *n.* 负荷，电荷；*v.* 充电
proportional *adj.* 比例的，成比例的
farad *n.* ［电］法拉（电容单位）

magnetic *adj.* 磁的，有磁性的

current-carrying wire in a circuit acts as an inductor! However, this type of "**stray**" inductance is typically **negligible**, just as we can usually ignore the stray resistance of our wires and only take into account the actual resistors. To store any appreciable amount of magnetic energy, one usually uses **a coil of** wire designed specifically to be an inductor. All the **loops**' contribution to the magnetic field add together to make a stronger field. Unlike capacitors and resistors, practical inductors are easy to make by hand. One can, for instance, **spool** some wire around a short wooden **dowel**, put the spool inside a plastic **aspirin** bottle with the leads hanging out, and fill the bottle with **epoxy** to make the whole thing **rugged**. An inductor like this, in the form of cylindrical coil of wire, is called a **solenoid**, as shown in Fig. 1. 4. Fig. 1. 5 is the symbol used to represent an inductor in a circuit regardless of its actual geometry.

stray *adj.* 偶遇的，零散的
negligible *adj.* 可以忽略的

a coil of 一卷
loop *n.* 环，线（绳）圈

spool *vt.* 缠绕在线轴上，缠绕
dowel *n.* 木钉
aspirin *n.* 阿司匹林（解热镇痛药），乙酰水杨酸
epoxy *adj.* 环氧的
rugged *adj.* 高低不平的，崎岖的，粗糙的
solenoid *n.* ［电］螺线管

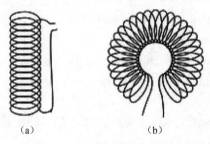

（a）　　　　　　（b）

Fig. 1. 4　Two common geometries for inductors

Fig. 1. 5　The symbol for an inductor

How much energy does an inductor store? The energy density is proportional to the square of the magnetic field strength, which is in turn proportional to the current flowing through the coiled wire, so the energy stored in the inductor must be proportional to I^2. We write $L/2$ for the constant of proportionality, giving

$$E_L = \frac{L}{2} I^2$$

As in the definition of inductance, we have a factor of $1/2$, which is purely a matter of definition. The quantity L is called the inductance of the inductor, and we see that its units must be **joules** per ampere squared. This **clumsy combination** of units is more commonly abbreviated as the **henry**, and the unit is H ($1\ H = 1\ J/A^2$).

joule *n.* ［物］焦耳
clumsy *adj.* 笨拙的
combination *n.* 结合，联合，合并
henry *n.* 亨利（电感单位）

Get to Work

I . Matching

Directions: *Match the following terms to appropriate definition or expression.*

1. capacitor or a capacitor   A. an electromotive force

2. microfarad   B. the unit of inductance

3. inductor   C. as an electric field to store electrical energy

4. henry   D. coil of wire with or without a magnetic core

5. emf   E. one millionth of a farad

II . Understanding Checking

Directions: *Mark Y (for YES) if the statement agrees with the information given in the text; N (for NO) if the statement contradicts the information given in the text.*

(　　) 1. Instead of batteries, we generally use capacitors and inductors to store energy in oscillating circuits.

(　　) 2. A simple capacitor consists of two metallic plants separated by a dielectric.

(　　) 3. Unlike capacitors and resistors, practical inductors are easy to make by hand.

(　　) 4. The amount of the stored energy of a charged capacitor is proportional to the applied voltage only.

(　　) 5. This clumsy combination of units, joules per ampere squared, is more commonly abbreviated as the henry.

III . Passage Completion

Directions: *Fill in each of the blanks with one of the words or expressions in the box, making changes if necessary.*

unlike	half	by	storage	two	

When two identical capacitances are placed in parallel, any charge deposited at the terminals of the combined double capacitor will divide itself evenly between the two parts. The electric fields surrounding each capacitor will be 1. _____ the intensity, and therefore store one quarter of the energy. Two capacitors, each storing one quarter the energy, give half the total energy storage. Since capacitance is inversely related to energy 2. _____, this implies that identical capacitances in parallel give double the capacitance. In general, capacitances are placed in parallel. This is 3. _____the behavior of inductors and resistors, for which series configurations give addition.

This is consistent with the fact that the capacitance of a single parallel-plate capacitor proportional to the area of the plates. If we have 4. _____ parallel-plate capacitors, and we combine them in parallel and bring them very close together side 5. _____ side, we have produced a single capacitor with plates of double the area, and it has approximately double the capacitance.

Ⅳ. Para. Translation

Directions: *Translate the follow paragraph from English into Chinese.*

If two inductors are placed in series, any current that passes through the combined double inductor must pass through both its parts. Thus, by the definition of inductance, the inductance is doubled as well. In general, inductances are placed in series, just like resistances. The same kind of reasoning also shows that the inductance of a solenoid is approximately proportional to its length, assuming the number of turns per unit length is kept constant.

Task Ⅱ Technology Application Understanding

1.3 Voltage Dividers

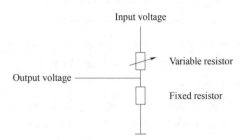

Learning Objectives

In this section, you will:

1. understand the function of voltage divider.

2. comprehend the use of a series dropping resistor R in Fig. 1.7, and the exact resistor value of R.

3. explain the disadvantage of the method provided in Fig. 1.7 in the actual application.

4. find out the values of resistors in Fig. 1.8.

Warm-up Activity

Any two-terminal linear network, composed of voltage sources, current sources, and resistors, can be replaced by an equivalent two-terminal network consisting of an independent voltage source in series with a resistor (Fig. 1.6).

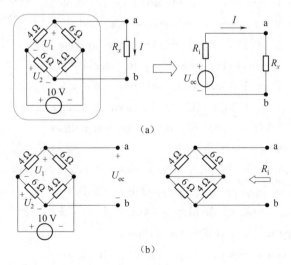

Fig. 1.6 Equivalent circuit model

(a) Compute the open circuit voltage, this is U_{oc}; (b) Compute the Thevenin resistance

Text

In many circuits, it is necessary to obtain a voltage not available from the main power source. Rather than have multiple power sources for all needed voltages, we can **derive** other voltages from the main power source. In most cases, the needed voltage is less than the voltage from the main source, so we can use resistors in an **appropriate configuration** to reduce the voltage from the power source, for use in a small circuit.

If we know precisely both the voltage and current required, we can simply connect a resistor **in series with** the power source, with a value **calculate**d **in accordance with** Ohm's Law. This resistor will drop some of the source voltage, leaving the right amount for the actual load, as shown in Fig. 1.7.

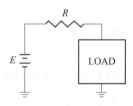

Fig. 1.7   Using a series dropping resistor

Usually, however, this doesn't work too well. The required value of the series **dropping resistor** will almost never be a standard value, and the cost of having special values **manufactured** for specific circuits is **prohibitive**. For example, suppose we have a 9 V battery as your main power source, and want to operate a load that requires 5 V at 3.5 mA. Our series resistor, R, must drop 4 V at 3.5 mA. Using Ohm's Law to calculate the required resistance value, we find that we need a resistance of $4/0.0035 = 1,142.8571\ \Omega$ or $1.1428571\ \mathrm{k\Omega}$. We have a choice between $1.1\ \mathrm{k\Omega}$ and $1.2\ \mathrm{k\Omega}$ as standard 5% values, but neither will give us what we want.

A more practical solution to the problem is to use two resistors in series, and use the voltage appearing across one of them. This configuration is known as a **voltage divider** because it divides the source voltage into two parts. The basic circuit is shown in Fig. 1.8.

In this circuit, the output voltage, U_{OUT}, can be set accurately as a **fraction** of the source voltage, E. Using our example above,

derive *vt.* 得自 *vi.* 起源

appropriate *adj.* 适当的
configuration *n.* 构造，结构，配置，外形

in series with 串联
calculate *v.* 计算
in accordance with 与……一致，依照

dropping resistor 降压电阻器

manufactured *adj.* 人造的
prohibitive *adj.* 禁止的，抑制的

voltage divider 分压器

fraction *n.* 小部分

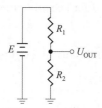

Fig. 1. 8　A voltage divider circuit

we want to select R_1 and R_2 such that we will drop 4 V across R_1, leaving 5 V across R_2. Since U_{OUT} is the voltage across R_2, this will give us the voltage we want.

We can get a U_{OUT} of 5 V, then, if we set $R_1 = 1.2$ kΩ and $R_2 = 1.5$ kΩ. We can also get the same U_{OUT} if we make $R_1 = 12$ kΩ and $R_2 = 15$ kΩ. The exact resistor values don't matter, so long as their ratio is correct.

The voltage divider is a very simple circuit that can be highly accurate if not **loaded down**.

load down 使负担过重；超载

 Get to Work

Ⅰ. Matching

Directions：*Match the following terms to appropriate definition or expression.*

1. the high-pass filter　　　　　A. a device that produces and converts electricity

2. oscillator　　　　　　　　　B. allows the signals above a certain frequency to pass

3. power supply　　　　　　　　C. the ability of a power supply to maintain a constant voltage

4. regulation　　　　　　　　　D. produce waveforms which may be sinusoidal, square, sawtooth etc.

5. rectifier circuit　　　　　　E. provide a unidirectional DC

Ⅱ. Understanding Checking

Directions：*Mark Y (for YES) if the statement agrees with the information given in the text; N (for NO) if the statement contradicts the information given in the text.*

(　　) 1. In most cases, the needed voltage is more than the voltage from the main source.

(　　) 2. If we know precisely both the voltage and current required, we can simply connect a resistor in parallel with the power source. This resistor will drop some of the source voltage.

(　　) 3. A more practical solution to the problem is to use two resistors in series, and use the voltage appearing across one of them. This configuration is known as a voltage divider.

(　　) 4. The voltage divider is a very simple circuit that can be highly accurate if loaded down.

(　　) 5. The required value of the series dropping resistor will almost never be a standard value, and the cost of having special values manufactured for specific circuits is prohibitive.

Ⅲ. Passage Completion

Directions: *Fill in each of the blanks with one of the words or expressions in the box, making changes if necessary.*

less	greater	dependent	same	consist

The voltage divider (Fig. 1.9) is the most basic analog input circuit. It 1. _____ of a variable resistor and a fixed resistor. The output voltage is 2. _____ on the ratio of the two resistors. For example, when the two resistance are the 3. _____, the output of voltage is half the input voltage. When the variable resistor is much 4. _____ than the fixed resistor, the output voltage is closer to the input voltage. When the variable resistor is much 5. _____ than the fixed resistor, the output voltage is closer to ground.

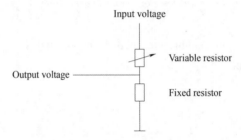

Fig. 1.9　Voltage divider

Ⅳ. Translation

Directions: *Translate the following sentences from Chinese into English.*

(1) 晶体管是电子技术中最重要的器件。

(2) 有三种基本的逻辑门，它们是与门、或门和非门。

(3) 1 V 电压施加在导体上产生了 1 A 电流，此时电阻为 1 Ω。

(4) 当把电动势加到线圈上时，线圈中就会产生感应电动势。

(5) 广义地说，把信息从一点传送到另一点就称为通信。

1.4　Filters

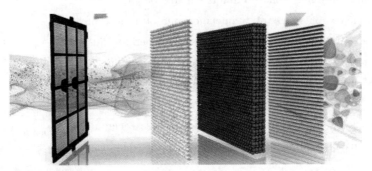

Learning Objectives

In this section, you will:

1. understand the function of a filter circuit.

2. learn about the differences between an active filter and a passive filter.

3. master the the four groups of filters in terms of their frequency response characteristics.

4. analyze how to use the filter in communications.

Warm-up Activity

Filters can be categorized in terms of their frequency response characteristics, as shown in Fig. 1.10.

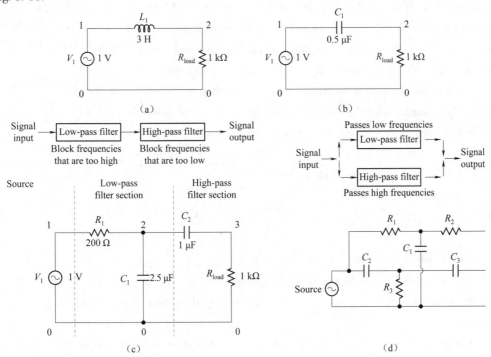

Fig. 1.10　Filters

Text

It is sometimes desirable to have circuits capable of selectively filtering one frequency or range of frequencies out of a mix of different frequencies in a circuit. A circuit designed to perform this frequency selection is called a **filter** circuit, or simply a filter. A filter is a **passive** or **active** circuit designed primarily **for the purposes of** modifying a signal or **source of power**. A passive filter requires no power for its **operation**, and always has a certain amount of loss. An active filter requires its own power, but may provide **gain**.

In a power supply, the filter is a network of capacitors, resistors and inductors. A good power supply filter will produce a smooth direct current output under variable load conditions, with good **regulation**.

Filters are used in **communications** practice to **eliminate** energy at some frequencies while allowing energy at other frequencies to pass with little or no **attenuation**. Such filters, if passive, are constructed using capacitors, inductor and sometimes resistors. Active filters generally utilize **operational amplifiers**.

Filters can be **categorized**, **in terms of** their frequency response characteristics, into four groups. The **band-pass filter** allows signal between two predetermined frequencies to pass, but attenuates all other frequencies. The **band-rejection** or band-stop **filter** eliminates all energy between two frequencies, and allows signals outside the limit frequencies to pass. The **high-pass filter** allows only the signals above a certain frequency to pass. The **low-pass filter** allows only the signal below a certain frequency to pass. The actual response characteristics of all four types of filter are highly variable, and depend on many different factors.

filter *n.* 滤波器	
passive *adj.* 无源的	
active *adj.* 有源的	
for the purpose of 为了	
source of power 电源	
operation *n.* 运转, 操作, 工作	
gain *n.* 增益	
regulation *n.* 稳压	
communication [计] 通信	
eliminate *vt.* 排除, 消除	
attenuation *n.* 衰减	
operational amplifier 运算 [操作] 放大器	
categorize *v.* 加以类别, 分类	
in terms of *adv.* 根据, 按照	
band-pass filter 带通滤波器	
band-rejection 带阻滤波器	
high-pass filter 高通滤波器	
low-pass filter 低通滤波器	

 Get to Work

Ⅰ. Matching

Directions: Match the following terms to appropriate definition or expression.

1. the high-pass filter   A. allows signal between two predetermined frequencies to pass

2. a filter   B. allows the signals above a certain frequency to pass

3. the low-pass filter   C. eliminates all energy between two frequencies

4. the band-pass filter

5. the band-rejection filter

D. allows the signals below a certain frequency to pass

E. a passive or an active circuit designed primarily for the purpose of modifying a signal or source of power

II. Understanding Checking

Directions: Mark Y (for YES) if the statement agrees with the information given in the text; N (for NO) if the statement contradicts the information given in the text.

(   ) 1. A passive filter requires power for its operation, and always has a certain amount of loss.

(   ) 2. In a power supply, the filter is a network of capacitors, resistors and inductors.

(   ) 3. Filters can be categorized, in terms of their frequency response characteristics, into four groups.

(   ) 4. The band-rejection filter allows signal between two predetermined frequencies to pass, but attenuates all other frequencies.

(   ) 5. The actual response characteristics of all four types of filter are highly variable, and depend on many different factors.

III. Passage Completion

Directions: Fill in the blanks with words according to the text.

1. A filter is a _____ or _____ circuit designed primarily for the purpose of modifying a signal or source of power.

2. In a power supply, the filter is a network of _____, _____ and _____ .

3. Filters can be categorized, in terms of their _____ characteristics, into four groups.

4. The high-pass filter allows only the signals _____ a certain frequency to pass.

IV. Translation

Directions: Translate the following sentences from Chinese into English.

1. 无源滤波器由电阻、电容和电感组成。

2. 有源滤波器由集成运放组成。

3. 滤波器按其频率特性可分为高通滤波器、低通滤波器、带通滤波器和带阻滤波器。

4. 滤波器主要用来滤除某些频率成分而允许另一些频率成分无衰减地通过。

5. 滤波器包括有源滤波器和无源滤波器。

Task Ⅲ Skill Honing

1.5 Features of Scientific English
科技英语的特点

科技英语不同于日常英语、文学英语等文体，它具有自身独有的规律和特点，主要体现在以下几个方面。

一、被动语句多

科技英语的目的是要阐述科学事实、描述发展变化、揭示本质特征等，在论述时应保持客观中立，强调准确性和可信度。因此尽量使用第三人称叙述，采用被动语态，避免主观色彩。例如：

The potential of fiber optics in other areas **has been realized**.

人们已经认识到光纤在其他领域的应用潜力。

（has been realized 是被动语态，翻译为"人们已经认识到"，转化为主动语态）

二、名词化结构多

名词化结构可以代替从句使语言更简练、结构更紧凑。例如：

A sound understanding of machine functions and performance is the basis for safe, accident-free and injury-free operation.

彻底了解机器的功能和性能是安全操作、无事故以及无伤害的前提。

（句中 a sound understanding of 是名词化结构）

三、非谓语动词多

大量使用分词、动名词、不定式、独立主格结构代替从句，也是为了缩短句子，使表达更加醒目直观、简明清晰。例如：

A sinusoidal oscillator generates a signal **having** a sine waveform.

正弦振荡器发出正弦波形的信号。

（having 是现在分词，作后置定语，代替定语从句）

The article deals with microwaves, **with particular attention paid** to radio location.

这篇文章是研究微波的，其中特别注意无线电定位问题。

（with…paid 是分词复合结构，在句中作补充说明）

The power station is automatic, **the sluices being worked** form the control board.

这个发电站是自动化的，闸门可以由控制板操纵。

（现在分词的被动形式，作独立主格结构，在句中作状语）

四、词类转换多

英语单词有不少词是一词多性，即既可以作名词，又可作动词、形容词、介词或副词。还有很多词是一词多义的，而且在不同的场合、不同的语境、不同的专业中意思也不相同。例如：

clear

adj. 明朗的 a clear day （晴天）

　　清楚的 a clear photograph （清晰的照片）

　　无罪的 a clear conscience （无愧的良心）

　　完全的 a clear victory （彻底的胜利）

　　无瑕疵的 a clear complexion （无瑕的肤色）

　　无障碍的 a clear view （一览无余的视野）

adv. 清楚地 speak clear （讲话清楚）

　　完全 He got clear away. （他逃得无影无踪。）

n. ［建］中空体内部的尺寸 The tube is 10 ft① wide in the clear. （这管子内径宽 10 ft。）

vt. 清除 clear the plates away （把盘子拿走）

　　跳过 The horse cleared the fence. （马跳过篱笆。）

　　净得 He cleared two thousand dollars last year. （去年他净赚了 2 000 美元。）

　　为船只结关 clear a ship at the customhouse （在海关办理船只的出港手续）

vi. 变明朗 The weather has cleared up. （天放晴了。）

五、短语结构多

Most of the products *on display* are new ones.

展出的产品大多数都是新产品。

（短语 on display 作定语）

This machine is *out of order*.

这台机器失灵了。

（短语 out of order 作表语）

六、复杂长句多

科技文章要求叙述准确，用词严谨，因此一句话里常常包含多个分句，这种复杂且长的句子居科技英语难点之首，在阅读翻译时要按照汉语习惯加以分析，以短代长，化难为易。例如：

An uncharged object contains a large number of atoms, each of which normally contains an equal number of electrons and protons, but with some electrons temporarily free from atoms.

不带电的物体含有大量的原子，每个原子通常含有等量的电子和质子，但是有一些电子暂时脱离了原子的束缚。

（这是一个由非限定性定语从句和 with + 名词 + 形容词的复合结构组成的长句。定语从句和复合结构对句子起补充说明作用。）

---

① 1 ft = 0.304 8 m。

Task Ⅳ　Job Task Challenges

1.6　Career Speaking：Job Interview

Ⅰ. Warm Up

Model

(A = Applicant　　I = Interviewer)

I：Mr. Zhang, I have read your resume. Why did you choose to major in mechanical engineering?

A：Many factors lead me to major in mechanical engineering. The most important factor is that I like *tinker*ing *with* machines.

I：What are you primarily interested in about mechanical engineering?

A：I like designing products, and one of my designs received an award. Moreover, I am familiar with CAD.

I：Why did you decide to apply for this position?

A：Your company has a very good reputation, and I am very interested in the field your company is in.

I：What do you think determines an employee's progress in a company such as ours?

A：Interpersonal and technical skills.

I：We have several applicants for this position. Why do you think you are the person we should choose?

A：I have the abilities, qualities and experience that you requested in your job advert, for example, I have three years' experience in designing products and I got leadership experience while serving the college student union as president.

I：That sounds very good. How do you see your career development?

A：After a few years of gaining experience in the company and furthering my professional qualifications I'd like to put my experience and skills to use in management. I want to become a supervisor in your R&D department.

Note

tinker with　　　　　　　　　　摆弄

Ⅱ. Matching

Directions：The following conversation is based on a job interview. "I" stands for an interviewer while "A" represents an applicant. Please match the corresponding question with its answer. Then, play roles with your partner.

I：Have you obtained any certificate of technical qualifications or license?

A：_____.

I：How many years have you had the driver license?

A：_____.

I：That's good. What special skills do you have, can you tell me?

A：_____.

I：What computer languages have you learned?

A：_____.

I：Have you gotten any special training in programming?

A：_____.

I：How do you think of your proficiency in written and spoken English?

A：_____.

S1：I have two years' driving experience.

S2：I have experience in computer operation, proficiency in Microsoft Windows, Microsoft Word and Microsoft Excel.

S3：I have learned English for 10 years, and I have passed College English Test Band 4 and 6. My spoken English is fairly good enough to express myself fluently.

S4：No, but I have database programming experience and network knowledge.

S5：Yes, I have received an Engineer Qualification Certificate and a driver license.

S6：Visual C + + and C.

III. Closed Conversation

Directions: The following conversation is based on a job interview. "I" stands for an interviewer while "A" represents an applicant. Please fill the information in the blanks according to the Chinese. Then, play roles with your partner.

I：What kind of person do you think you are?

A：_____（我觉得自己精力充沛，做事有热情）.

I：What are your weaknesses?

A：_____（有时为了把事情办得完美些，我让自己背上太多的压力）.

I：How would your friends or colleagues describe you?

A：_____（他们认为我是很友好、敏感、关心他人和有决心的人）.

I：What personality do you admire?

A：_____（我欣赏诚实、灵活而且容易相处的人）.

I：How do you spend your leisure time?

A：_____（我爱玩游戏和体育运动）.

I：What qualities would you expect of persons working as a team?

A：To work in a team, in my opinion, _____
(合作精神和进取精神两者皆不可少).

Ⅳ. Semi-open Conversation

Directions：The following conversation is based on a job interview. "I" stands for an interviewer while "A" represents an applicant. Please fill in the blanks according to your own information. Then, play roles with your partner.

I：Which school are you attending?

A：_____.

I：When will you graduate from that university?

A：_____.

I：What is your major?

A：_____.

I：How did you get on with your studies in university?

A：_____.

I：Which course do you like best?

A：_____.

I：Did you get any honors and awards at college?

A：_____.

Ⅴ. Live Show

Challenge 1

Directions：You, a new graduate from a vocational college, are hunting a job that is related to your major. Please act it out with your partner or partners. The model and the useful expressions below are just for your reference.

Model

(A = Applicant   I = Interviewer)

A：Good morning. I came in answer to your advertisement for a salesperson.

I_1：I see. What experience have you had?

A：I'm afraid I haven't had any experience in just this sort of work.

I_2：Have you got any selling experience at all?

A：I worked in a department store in a small town last summer vacation.

I_1：Now, tell us about your educational background.

A：I graduated from Hezhe Vocational College. I majored in Optoelectronic Technology.

I_2：What courses have you completed?

A：The courses I completed are Circuits, C Language, Program Principle and Application of Microcontroller, Laser Principles, Photoelectric Detection Technology, Optical Fiber Communication and so on.

I_1：Why would you like to work with us?

A：It's an interesting job, and your company is one of the best known. Although I have no work experience as a salesperson, the job description you sent was very interesting.

I$_2$: Do you know anything about this company?

A: Yes, a little. As far as I know, your company is a world famous one which produces optical devices.

Notes

Analog Circuits	模拟电路
Digital Circuits	数字电路
Circuits	电路
Signals & Systems	信号与系统
Electromagnetic Field (EMF)	电磁场
Digital Signal Processing (DSP)	数字信号处理
Optical Fiber Communication (OFC)	光纤通信
Data Communication	数据通信
Mobile Communication	移动通信
Modern Switching Technology	现代交换技术
Communication System Principle	通信系统原理
Communication Electro Circuit	通信电子线路
The C Programming Language	C 语言
Database Principles	数据库原理
MATLAB	
Software Technique	软件技术
Embedded System Design	嵌入式系统设计
Codesign of Embedded Systems	嵌入式系统协同设计
Principles of Microcomputer	微机原理
Principles and Applications of Microcontroller	单片机原理与应用
Principles and Applications of DSP	DSP 原理与应用
Computer Network	计算机网络
Laser Principles	激光原理
Photoelectric Detection Technology	光电探测技术
Principles of Optics	光学原理
Randomized Procedure	随机过程
Infrared System	红外系统

Challenge 2

Directions: You, an employee in an Industry Laser company, are not satisfied with your current salary and benefits and decide to talk to your boss. Please act it out with your partner or partners. The model and the useful expressions below are just for your reference.

Useful Expressions

Why do you think you are qualified for this position?

I have worked at CBA Company for 4 years and got a lot of practical experience.

What's your technical post title now?

I am a senior mechanical design engineer.

What made you decide to change your job?

Because I want to change my working environment and seek new challenges.

What do you think is the most important qualification for a salesperson?

I think it is confidence in himself and his products.

Do you have any questions to ask about this job?

I'd like to know something about the salary.

1.7　Writing：Resume

求职简历是求职者生活、学习经历及职位期望等的简述，一般以大纲或一览表的形式表示，其目的是让招聘单位尽快地了解求职者的基本情况。英文简历包括求职目标、个人资料、教育经历、工作经历、所获奖励和社会活动、个人爱好或特长等。以下给出两份示例。

Sample 1

PERSOANL DATA

Name：Edward Brown　　　　Sex：Male　　　　Date of Birth：Aug. 20，1985

Height：178 cm　　　　　　Weight：70 kg　　　 Health：Excellent

Address：32 Gilmore Lane，Boston，Massachusetts 02101

Telephone No. ：（617）555-1212　　E-mail：Whiteman@ yahoo. com

JOB OBJECTIVE

Director of Banking Training

EDUCATION

West Springfield University, Illinois

Master of Business（MBA），June 2009

West Springfield University, Illinois

Bachelor of Finance，June 2006

WORK EXPERIENCE

August 2008—Present　　　Standard Chartered Bank，Springfield，Illinois Data Services

Manager/Administrative Assistant Supervise，manage and

train 22-person office staff

Prepare for meetings and correspond with member

representatives on upcoming meetings.

June 2006—July 2008	Standard Chartered Bank, Springfield, Illinois
	Bookkeeper
	Received cash and check receipts, maintained ledger book and computer record of band deposits.
	Wrote and distributed employee and contractor checks.

SKILLS

Microsoft Word 2007

Microsoft Excel 2007

HONORS

Model Worker in Standard Chartered Bank (Springfield Branch) in 2010

West Springfield University Trustee Scholarship Recipient in 2008

HOBBIES

Cycling, traveling, fishing

References   Provided upon request

Sample 2

Full Name:	Li Jing
Address:	No. 113 Wuda Road, Ximing City
Mobile:	167-0078-1148
Date of birth:	March 5, 1990
Sex:	Female
Nationality:	Chinese
Marital Status:	Single
Health:	Excellent
Job Objective:	An technician in a laser company with opportunity for advancement
Education:	
	2008—2011 Changshan Vocational College
	Major: Laser Processing
	2005—2008 No. 2 High School, Hushui County
Part-time Work Experience:	
	July 2009—August 2009 A salesperson in Jingming Computer Company
	July 2010—August 2010 A worker operating laser cutting in Feida Laser Co., Ltd.

Awards & Honors:	The National Scholarship Recipient in 2010
	"Excellent Student" of Changshan Vocational College in 2009
Strengths:	Being good at communication and cooperation
	Excellent team spirit and professional ethics
Interests:	Watching movies, filming and listening to music
References:	Available upon request

Challenge

Directions: Finish a resume for yourself in terms of your own conditions and situations.

Task V　Output and Evaluation

Basic Circuit Concepts					
Target	Understand and describe a basic circuit				
Requirement	Make an English presentation according to lesson 1. 1. Draw a mind map. Produce Micro-video to express professional knowledge. Add accurate Chinese and English subtitles to the video.				
Contents	Basic Circuit Concepts： A basic DC circuit Characteristics Relationship				
Group：_____	**Project**	**Name**	**Software**	**Score**	**Requirements**
	Mind map				The logic should be sound, and the keywords used should be accurate.
	Script				The format should adhere to the specified guidelines, and it should cover all necessary contents.
	PPT				It should be consistent with the logic presented in the mind map. Clear and concise content/Consistent Formatting/Limited text/Engaging graphics and animations.
	Speaking				Clear/fluent
	Subtitle				They should be bilingual, correct and synchronized with the student's speech.
	MP4				The video should effectively convey knowledge, be accurate, and visually engaging.
Operating environment	Win7/Win8/Win9/Win10/Win11/Mobile phone				
Product features					

Basic Circuit Concepts
思维导图

Basic Circuit Concepts
微课

27

Script reference

Basic Circuit Concepts

标题	专业内容（讲稿与中英字幕）	PPT 中需显示的文字	表现形式	素材
每个知识点的要点，或者小节名字	做报告时候的英文讲稿 视频中的英文字幕	需要特别提醒的文字	讲解到某个地方或某句话时，出现的图片、动画或者文字	照片，或者视频
Contents	**Basic circuit concepts** Contents： 1. A basic DC circuit 2. Characteristics 3. Relationship **基本电路概念** 内容： 基本直流电路 特性 关系	**Basic circuit concepts** Contents： 1. A basic DC circuit 2. Characteristics 3. Relationship	文字	
	Fig. 1 shows the basic type of an electrical circuit, in the form of a block diagram. It consists of a source of electrical energy, some sort of load to make use of that energy, and electrical conductors connecting the source and the load. 右图 1 以框图的形式展示了一个基本电路。它包含电源，使用电源的负载，和连接电源和负载的导线。	Source Load Conductors	图示	图 1　基本电路框图 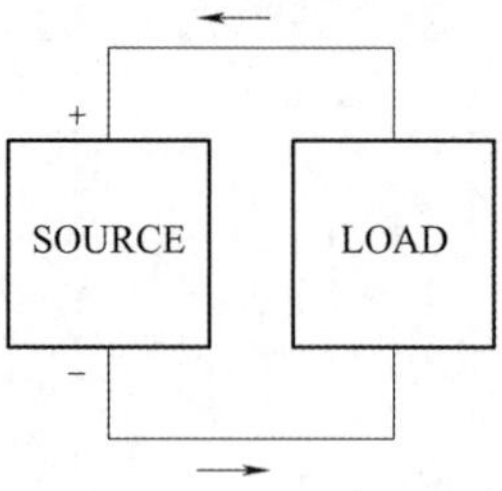 + SOURCE　　LOAD −
1. A basic DC circuit	The electrical source has two terminals, designated positive （＋） and negative （－）. As long as there is an unbroken connection from source to load and back again as shown here, electrons will be pushed from the negative terminal of the source, through the load, and then back to the positive terminal of the source. The arrows show the direction of electron current flow through this circuit. Because the electrons are always moving in the same direction through the circuit, their motion is known as a direct current. 电源有两极，分别为正极 （＋） 和负极 （－）。只要从电源到负载的连接不被破坏，电子就会从电源的负极流经负载，然后又返回电源的正极。箭头显示了电子流经电路的方向。因为电子在电路中沿相同方向移动，它们的运动被称为直流电。	Two terminals An unbroken connection The arrows	图示	图 1　基本电路框图

续表

标题	专业内容（讲稿与中英字幕）	PPT 中需显示的文字	表现形式	素材
1. A basic DC circuit	The source can be any source of electrical energy. In practice，there are three general possibilities：it can be a battery, an electrical generator, or some sort of electronic power supplies. 电源是任何电的来源，通常有三种基本类型，它可以是电池、发电机，或者是某种电子电源。	**Source** three general possibilities： Battery； Electrical generator； Electron power supplies	图示	图2　电池  图3　发电机 
	The load is any device or circuit powered by electricity. It can be as simple as a light bulb or as complex as a modern high-speed computer. 负载可以是任何器件或由电能供电的电路。它可以是简单的灯泡（图4），也可以是复杂的现代高速计算机（图5）。	Load： Light bulb； Modern high-speed computer	图示	图4　灯泡 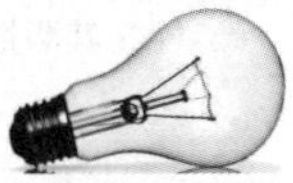 图5　计算机 
2. Characteristics	The electricity provided by the source has two basic characteristics，called voltage and current. 由电源提供的电能有两个基本特性，称为电压和电流。		文字	图6　电压  图7　电流 
	Voltage The electrical "pressure" that causes free electrons to travel through an electrical circuit. Also known as electromotive force. It is measured in volts. 电压即电学"压力"，它促使电子流过电路，也称为电动势，其测量单位是伏特。	Voltage	定义	图8　公式 $E = I \times R$ 图6　电压

标题	专业内容（讲稿与中英字幕）	PPT 中需显示的文字	表现形式	素材
2. Charact-eristics	**Current** The amount of electrical charge（the number of free electrons）moving past a given point in an electrical circuit per unit of time. Current is measured in amperes. 电流是单位时间内流过电路指定点的电子数量，其测量单位是安培。	Current	定义	图8 公式 图7 电流
	Resistance That characteristic of a medium which opposes the flow of electrical current through itself. Resistance is measured in ohms. 电阻是介质阻碍电流流过它自身的特性，其测量单位是欧姆。	Resistance	定义	图8 公式 图9 电阻 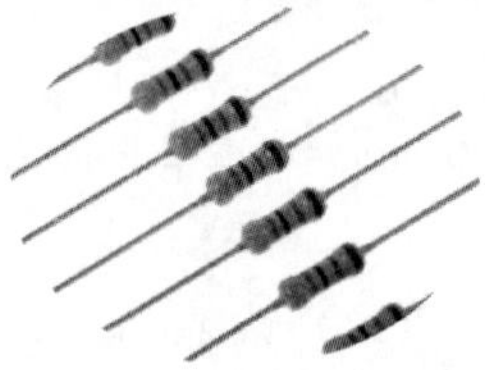
3. Relation-ship	The relationship between voltage，current，and resistance in an electrical circuit is fundamental to the operation of any circuit or device. 在电路中电压、电流和电阻的关系是任何电路和设备工作的基础。	The relationship between voltage，current，and resistance in an electrical circuit is fundamental to the operation of any circuit or device.	文字	
	Verbally，the amount of current flowing through a circuit is directly proportional to the applied voltage and inversely proportional to the circuit resistance. 通常，流经电路的电流与电压成正比，与电阻成反比。	Verbally Directly proportional Inversely proportional		图10 Directly proportional 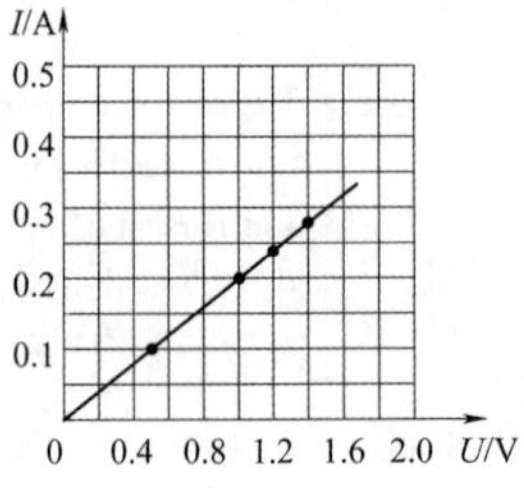 图11 Inversely proportional 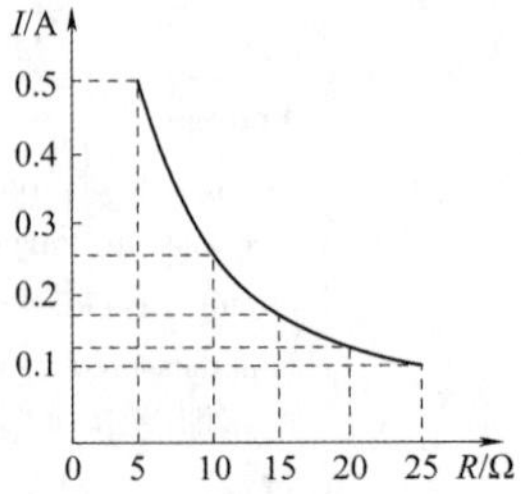

续表

标题	专业内容（讲稿与中英字幕）	PPT 中需显示的文字	表现形式	素材
3. Relation-ship	By explicit definition, one volt of electrical pressure can push one ampere of current through one ohm of resistance. Two volts can either push one ampere through a resistance of two ohms, or can push two amperes through one ohm. 准确的定义，一伏特的电压能使一安培的电流流过一欧姆的电阻。两伏特的电压能使一安培的电流流过两欧姆的电阻，或者使两安培的电流流过一欧姆的电阻。	Explicit definition		
	Mathematically, $$E = I \times R$$ where E is the applied voltage, or EMF; I is the circuit current; R is the resistance in the circuit. 数学表达式如下： $$E = I \times R$$ 式中，E = 电压或电动势；I = 电流； R = 电阻。	Mathematically	图文	图 8　公式
	The end. Thank you!	The end. Thank you!		

Task Ⅵ   Knowledge Expansion

1.8   Ohm's Law

One thing we need to be able to do when we see a schematic circuit diagram is to perform mathematical calculations to define the precise behavior of the circuit. All information required to perform such calculations should be included on the schematic diagram itself. That way, the information is all in one place, and any required detail can be determined readily.

Consider the basic circuit shown in Fig. 1.11. We know immediately that the battery voltage is 6 V and that the resistor is rated at 1,000. Now, how can we determine how much current is flowing through this circuit?

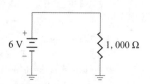

Fig. 1.11   Basic circuit

If you go back to the basic circuit, you'll note that the relationship between voltage, current, and resistance is given as $E = IR$. Using basic algebra we can also rewrite this as:

$$R = E/I$$
$$I = E/R$$

These three equations describe Ohm's Law, which defines this relationship.

In the circuit shown in Fig. 1.11, we see that $E = 6$ V and $R = 1,000$ Ω. To find the current flowing in this circuit, we must select the equation that solves for I. Using that equation, we note that:

$$I = E/R$$
$$I = 6(V)/1,000(Ω)$$
$$I = 0.006(A) = 6(mA)$$

All calculations involving Ohm's Law are handled in exactly the same way. If the circuit gets complex, the calculations must be tailored to match. However, each calculation is still just as simple as this.

Exercises

Directions: Choose the best answers from the four choices.

1. The flow of electrons through a conductor is called _____ .

A. voltage   B. current

C. resistant   D. Ohm's Law

2. The relationship between volts, amperes and ohms can be represented by _____ .

A. voltage   B. current

C. resistant   D. Ohm's Law

1.9　What Is Electricity?

The modern science of electricity originated with Benjamin Franklin, who began studying and experimenting with it in 1747. In the course of his experiments, Franklin determined that electricity was a single force, with positive and negative aspects. Up to that point, the prevailing theory was that there were two kinds of electricity, positive and negative.

To describe his experiments and results, Benjamin Franklin also coined some twenty five new terms, including armature, battery, and conductor. His famous kite-flying experiment in a thunderstorm was performed in 1752, near the end of his work in this field.

Since then, many scientists, inventors, and entrepreneurs around the world have performed their own experiments, verifying and building on Benjamin Franklin's beginnings in the field. Now, some 250 years later, we use electricity in almost every aspect of our daily lives. In some cases, we may not even realize that electricity is involved as an integral part of our activities.

So just what is electricity? Let's start with the dictionary definition to give all of us some common ground. The American Heritage Dictionary actually gives four specific definitions as below.

1. The class of physical phenomena arising from the existence and interactions of electric charge.

2. The physical science of such phenomena.

3. Electric current used or regarded as a source of power.

4. Intense emotional excitement.

We will skip the fourth definition as having no useful connection to the other three, and deal with electricity as a physical phenomenon which may be studied and manipulated using the tools of science.

When Benjamin Franklin developed his hypotheses about electricity, he arbitrarily assumed that the actual carriers of electrical current had a positive electrical charge. All of his theories, calculations, and descriptions were based on this assumption. Fortunately, his experiments still worked even with this incorrect assumption built into them. This "conventional" assumption was used for 200 years or more, and is still built into many of the common rules and procedures used to design and analyze electrical devices and behaviors.

We now know that the actual carriers of electricity are electrons, which have a negative electrical charge as defined in our system of science. Because of this, "electron theory" has been replacing "conventional theory" in schools and in regular usage.

The obvious next questions are:

1. What is an electron?

2. Where do electrons come from?

3. How do they carry an electrical charge from place to place?

We'll begin answering those questions when we take a closer look at electrons.

Exercise

Directions: Answer the following question.

Can you tell us what is electricity in your opinion?

Unit 2   Optics

In this unit, you are going to learn the following contents:

Task Ⅰ   Technical Principles and Device Cognition

2. 1   Nature of Light

2. 2   Reflection and Refraction

Task Ⅱ   Technology Application Understanding

2. 3   Numerical Aperture（NA）

2. 4   Electromagnetic Radiation

Task Ⅲ   Skill Honing

2. 5   Translation of Words and Phrases

Task Ⅳ   Job Task Challenges

2. 6   Career Speaking: Meeting Guests

2. 7   Writing: Letter of Thanks

Task Ⅴ   Output and Evaluation

Task Ⅵ   Knowledge Expansion

2. 8   What Is light

2. 9   Convex Lens

Unit 2

Optics

Task I   Technical Principles and Device Cognition

2.1   Nature of Light

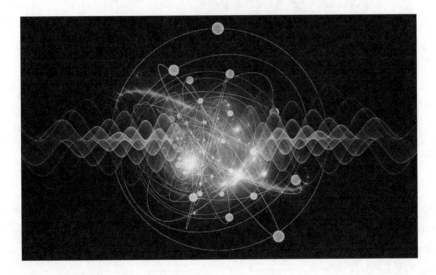

Learning Objectives

In this section, you will:

1. identify the range of visible wavelengths.

2. learn about the relationship between light and radio waves.

3. comprehend the electron volts.

4. learn about wave nature of light.

5. explain particle theory.

6. translate the example of this article and express it in three parts: known, find, and solution.

7. describe what phenomenon the particle theory can explain.

Warm-up Activity

Do you know the names of the apparatus in Fig. 2. 1?

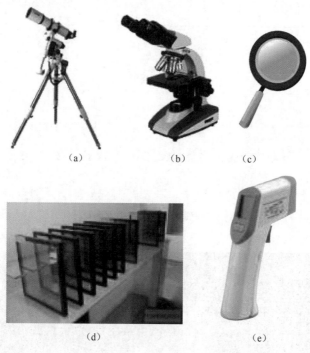

(a)　　　　(b)　　　　(c)

(d)　　　　(e)

Fig. 2. 1　Apparatus

 Text

Although light **pervades** human existence, its fundamental nature remains at least a partial mystery. We know how to **quantify** light phenomena and make predictions, based on this knowledge, and we know how to use and control light for our own convenience. Yet light is often interpreted in different ways to explain different experiments and observations: sometimes light behaves as a **wave**, sometimes light behaves as a **particle**.

Wave Nature of Light

Many light phenomena can be explained if we look at light as being an electromagnetic wave having a very high **oscillation** frequency and a very short wavelength. We use the **term optic** (as well as the term light) to refer to frequencies in the infrared, visible, and ultraviolet portions of the spectrum. We do this because so many of the same analyses, techniques, and devices are applicable to these ranges.

The range of frequencies (or wavelengths) that primarily

pervade v. 遍及
quantify v. 量化

wave n. 波
particle n. 粒子

oscillation n. 摆动，振动

term n. 术语
optic adj. 眼的，视觉的，光学上的

interests us is introduced here. Visible wavelengths extend from 0. 4 μm (which we distinguish as the color blue) to 0. 7 μm (which appears to us as red). **Silica** glass fibers are not very good transmitters of light in the visible region. They attenuate the waves to such an extent that only short transmission links are practical. Losses in the ultraviolet are even greater. In the infrared, however, these regions in which glass fibers are relatively efficient transmitters of light. These regions occur at wavelengths around 0. 85 μm and between 1. 26 μm and 1. 75 μm. These wavelengths are sometimes referred to as the fiber transmission windows.

silica *n.* ［化］硅石

Although light waves have much higher frequencies than **radio waves**, they both obey the same laws and share many characteristics. All electromagnetic waves have electric and magnetic fields associated with them, and they all travel very quickly.

radio wave 无线电波

In empty space (usually referred to as free space), electromagnetic waves travel at the **velocity** 3×10^8 m/s. This velocity, indicated by c, is appropriate for wave travel in the atmosphere. In solid media, the wave velocity differs. Its value depends on the material and on the geometry of any wave guiding structure that is present. The wavelength of a light beam is given by

velocity *n.* 速度，速率，迅速，周转率

$$\lambda = v/f \tag{2.1}$$

where v is the **beam** velocity and f is its frequency. The frequency is determined by the emitting source and does not change when the light travels from one material to another. Instead, the velocity difference causes a change in wavelength according to Eq. (2.1).

beam *n.* 梁，桁条，（光线的）束

As an example, consider radiation at 0. 8 μm. Using Eq. (2.1), with $v = c$, yields the frequency 3.75×10^{14} Hz. This is a fast oscillation indeed. The period of this oscillation (the reciprocal of its frequency) is then 2.67×10^{-15} s, an extremely short time span. We should also note that the wavelengths of optic beams are of the order of 1 μm near the visible spectrum. Optic wavelengths are so small that most devices used in a fiber system have dimensions of many wavelengths. This is unlike the situation at lower frequencies, where device sizes can be a wavelength or less.

Particle Nature of Light

So far, we have looked at light as being a wave. Sometimes light behaves unlike a wave and instead behaves as though it were made up of very small particles called **photons**. The energy of a single photon is

photon *n.* ［物］光子

$$W_p = hf \qquad (2.2)$$

where $h = 6.626 \times 10^{-34}$ J·s and is called Planck's constant. The energy computed from Eq. (2.2) has the units joules. It is impossible to break light into divisions smaller than the photon. Ordinarily, beams of light contain huge numbers of photons. The following example illustrates this.

Example

Find the number of photons incident on a detector in 1 s if the optic power is 1 μW and the wavelength is 0.8 μm

Solution From Eq. (2.1) and Eq. (2.2), the energy of a single 0.8 μm photon is

$$W_p = hf = hc/\lambda = 2.48 \times 10^{-19} \text{ J}$$

Because power is the rate at which energy is delivered, we can write the total energy

$$W = Pt$$

Multiplying the power, 1 μW, times the 1 s time interval yields the energy 1 μJ.

The number of photons required to make up 1 μJ is

$$\frac{W}{W_p} = \frac{10^{-6}}{2.48 \times 10^{-19}}$$
$$= 4.03 \times 10^{12}$$

In example above, if we reduce our observation time to as little as 1 ns, we will still receive over 4,000 photons. The most sensitive receivers can detect the presence of radiation when only a few photons arrive.

A convenient unit of energy for photons is the electron volt. It is the kinetic energy, acquired by an electron when it is accelerated by 1 V of potential difference. The relationship between electron volts and joules is

$$1 \text{ eV} = 1.6 \times 10^{-19} \text{ J}$$

The 0.8 μm photon treated in example has an energy in electron volts given by

$$2.48 \times 10^{-19} \text{J}/(1.6 \times 10^{-19} \text{ J/eV}) = 1.55 \text{ eV}$$

Particle theory explains generation of light by sources, such as **light emitting diodes**, lasers, and **laser diodes**. It also explains detection of light by conversion of optic radiation to electrical current.

light emitting diodes 发光二极管（缩写为 LED）

laser diodes 激光二极管

Get to Work

I. Matching

Directions: Match the words or expressions in the left column with the Chinese equivalents in the right column.

1. radio waves	A. 电磁波
2. oscillation frequency	B. 振荡频率
3. photon	C. 焦耳
4. electromagnetic wave	D. 光子
5. joule	E. 无线电波

II. Understanding Checking

Directions: Mark Y (for YES) if the statement agrees with the information given in the text; N (for NO) if the statement contradicts the information given in the text.

(　　) 1. Sometimes light behaves as a wave, sometimes light behaves as a particle.

(　　) 2. Many light phenomena can be explained if we look at light as being an electromagnetic wave having a very high oscillation frequency and a very short wavelength.

(　　) 3. Electromagnetic waves travel at the velocity 3×10^8 m/s all the time.

(　　) 4. A convenient unit of energy for photons is the electron volt.

(　　) 5. The relationship between electron volts and joules is 1 eV = 1.6×10^{-19} J.

III. Passage Completion

Directions: Fill in each of the blanks with one of the words or expressions in the box, making changes if necessary.

of	with	assumption	nature	that

For a long time, physicists tried to explain away the problems with the classical theory of light as arising from an imperfect understanding of atoms and the interaction of light with individual atoms and molecules. The ozone paradox, for example, could have been attributed to the incorrect 1. _____ that the ozone layer was a smooth, continuous substance, when in reality it was made 2. _____ individual ozone molecules. It wasn't until 1905 that Albert Einstein threw down the gauntlet, proposing 3. _____ the problem had nothing to do 4. _____ the details of light's interaction with atoms and everything to do with the fundamental of 5. _____ light itself.

IV. Translation

Directions: Translate the following sentences from Chinese into English.

1. 可见光的波长范围是 $0.4 \sim 0.7$ μm。
2. 光波同时具有波和粒子的双重性质。
3. 真空中电磁波的速度是 3×10^3 m/s。
4. 通常光束是由大量光子组成的。
5. 光子的一个便于计算的能量单位是电子伏特。

2.2　Reflection and Refraction

Learning Objectives

In this section, you will:

1. learn about the definition of incident light, reflected light.

2. learn about angle of incidence, angle of reflection.

3. master the relationship between angle of incidence and the angle of reflection.

4. explain the refraction phenomenon when light travels from air into water.

Warm-up Activity

Please enjoy the scenery in Fig. 2.2, and share your own.

Fig. 2.2　Scenery

Text

In talking about the fundamental nature of light, we **indicated** that light tends to travel in a straight line, unless it is acted on by some external force or condition. The obvious next question is, "What kinds of forces or conditions can affect light, and how?"

To answer this question, we start with what we can see in our daily life. For example, we already know that light won't pass through the wall of a house, but it will go through a window. **Furthermore**, some windows **introduce** noticeable **distortion** in what we see. Looking carefully at the glass of such a window, we can see that it is uneven, perhaps with **ripples** across its surface.

We have also seen that some surfaces show accurate **images** of an actual **scene** nearby, while other surfaces show distorted images, but most surfaces only show one or more colors no matter how we look at them.

On top of that, some surfaces seem very much darker than their surroundings, while other surfaces seem just as bright as their surroundings. Indeed, occasionally you can see something that looks brighter than its surroundings, both at night and in broad daylight.

A. Reflection

When light reflects off a surface, it follows some rather basic rules which have been gradually determined by observation. A ray of light **approaches** a reflecting **horizontal** surface at an angle of 45°, bounces from the surface, and leaves at an angle of 45°.

So that we can agree fully on what we are talking about, we need to define a few terms.

Incident Light

Light approaching a surface is known as incident light. This is the incoming light before it has reached the surface.

Reflected Light

After light has struck a surface and **bounced** off, it is known as reflected light. This is the light that is now departing from the surface, as shown in Fig. 2.3.

Fig. 2.3   **Reflected light**

---

indicate *vt.* 指出，显示，象征，预示

furthermore *adv.* 此外，而且
introduce *vt.* 传入，引进
distortion *n.* 扭曲，变形，曲解，失真
ripple *n.* 波纹
image *n.* 图像
scene *n.* 现场，场面，情景，景色

reflection *n.* 反射

approach *vi.* 靠近
horizontal *adj* 水平的

incident light 入射光

reflected light 反射光

bounce *v.* （使）反跳，弹起

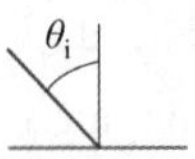

Angle of Incidence

The angle at which a ray of light approaches a surface, reflective or not, is called *the angle of incidence* (Fig. 2. 4). It is measured from an imaginary line perpendicular to the plane of the surface in question to the incoming ray of light.

Fig. 2. 4　Angle of Incidence

Angle of Reflection

Once the light has reflected from a reflective surface, the angle at which the light departs from the surface is called *the angle of reflection*. This angle is also measured from a **perpendicular** to the reflecting surface to the departing ray of light.

perpendicular *n.* 垂线；*adj.* 垂直的，正交的

When light reflects from a surface, the angle of reflection is always equal to the angle of incidence.

B. Refraction

Refraction is the name given to the observed phenomenon that light changes direction, or "bends", as it passes the **boundary** between one **medium** and another. This is shown in Fig. 2. 5, in a general sense.

boundary *n.* 边界，分界线

medium *n.* 媒介

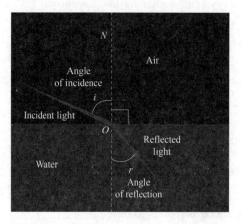

Fig. 2. 5　Refraction

Refraction 动画

Here, we see a beam of light traveling through air, until it meets a pool of water. It arrives at some angle to the surface, as shown in Fig. 2. 5. As it passes through the boundary, going from air into water, it actually slows down. Since even a single ray of light has a **finite** thickness, the part that enters the water first slows down

finite *adj.* 有限的，[数] 有穷的，限定的

first, causing the light ray to change direction to a steeper angle in the water.

If we change the angle at which the light enters the water, we find that the angle of the light in the water also changes, such that we see no change at all if the light source is directly overhead so that the entering ray of light is perpendicular to (in mathematical terms normal to) the surface. As we change the entering angle more and more away from the perpendicular, we see that the ray of light in the water has bent more and more away from the direction taken by that ray of light in the air.

 Get to Work

I. Matching

Directions: Match the words or expressions in the left column with the Chinese equivalents in the right column.

1. incident light	A. 折射
2. reflected light	B. 入射光
3. refraction	C. 水平的
4. horizontal	D. 反射光
5. perpendicular	E. 垂直的

II. Understanding Checking

Directions: Mark Y (for YES) if the statement agrees with the information given in the text; N (for NO) if the statement contradicts the information given in the text.

(　　) 1. In talking about the fundamental nature of light, we indicated that light tends to travel in a straight line, unless it is acted on by some external force or condition.

(　　) 2. Incident light is the incoming light before it has reached the surface.

(　　) 3. Light approaching a surface is known as reflected light.

(　　) 4. Both angle of reflection and angle of incidence are measured from a line perpendicular to the surface.

(　　) 5. Light can pass through the wall of house.

III. Passage Completion

Directions: Fill in each of the blanks with one of the words or expressions in the box, making changes if necessary.

refraction	process	although	various	different
with	psychological	definition	category	for example

Physical optics is concerned with the creation, nature, and properties of light. 1. _____ optics pertains to the role of light in vision. Geometrical optics deals 2. _____ the properties of reflection and 3. _____ of light, as part of the study of mirrors, lenses, and optical fibers.

As you can see from the dictionary 4. _____ above, the study of optics falls into three general 5. _____: light itself, what it is and how it behaves; How we perceive light through the sense of sight; and how light can be manipulated through such 6. _____ as reflection and refraction. 7. _____ we talk of these categories as if they were separate and distinct, actually the whole reality of sight and vision constantly involves all of these categories in 8. _____ combinations.

9. _____, consider a rainbow. This visual phenomenon not only involves both reflection and refraction, but also is a practical demonstration that different colors of light refract, 10. _____ even under fixed circumstances.

Ⅳ. Translation

Directions: Translate the following sentences from Chinese into English.

1. 当光线从表面反射，它遵循一些由观察得到的基本规则。

2. 光到达一个表面然后反射出来的光被称为反射光。

3. 光从一种物质到另一种物质，在界面处光的方向发生改变，或"弯曲"的现象称为折射。

4. 通常光束是由大量光子组成的。

5. 当入射光越来越远离垂线时，我们可以看到光在水中的方向越来越偏离入射光在空气中的方向。

Task Ⅱ　Technology Application Understanding

2.3　Numerical Aperture （NA）

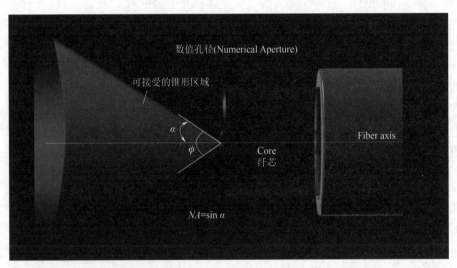

Learning Objectives

In this section, you will:

1. understand meaning of numerical aperture.

2. learn about a major advantage of the graded-index multimode fibers.

3. comprehend the calculating equation of NA.

Warm-up Activity

Please describe the condition of total reflection, as shown in Fig. 2.6.

Phrases for reference: total reflection; a vacuum, index of refraction ($n = 1$); the up media with index of refraction n_1, the media below with index of refraction n_2.

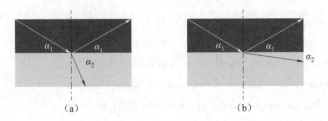

Fig. 2.6　Total reflection

（a）$n_2 > n_1$; （b）$n_2 < n_1$

Text

In optics, the numerical aperture of an optical system is a dimensionless number that characterizes the range of angles over which the system can accept or emit light. By incorporating index of refraction in its definition, NA has the property that it is constant for a beam as it goes from one material to another, provided there is no optical power at the interface. The exact definition of the term varies slightly between different areas of optics. Numerical aperture is commonly used in microscopy to describe the acceptance cone of an objective (and hence its light-gathering ability and resolution), and in fiber optics, in which it describes the range of angles within which light that is incident on the fiber will be transmitted along it.

Numerical aperture is a **critical performance specification** for multimode fibers. It indicates the maximum angle at which a **particular** fiber can accept the light that will be transmitted through it. The higher an optical fiber's NA, the larger the cone of light that can be coupled into its core.

Graded-index multimode fibers have a large NA. This is a major advantage of the product: it enables them to be used with relatively low-cost optical components and light sources such as light emitting diodes (LEDs) and **vertical cavity surface emitting lasers** (VCSELs). LEDs and VCSELs, which have large spot sizes, can be easily coupled to multimode fibers. In contrast, single-mode fibers, which have a small NA, typically use narrow width lasers as power sources and carry only one mode of light straight through a very narrow core. Transmitter **alignment** and **tolerances** must be very precise to couple the small beam of light into the tiny core of a single-mode fiber. This drives up the cost of single-mode components.

Multimode fibers allow more modes of light to be transmitted, resulting in greater pulse spreading, or dispersion, and less bandwidth. Consequently, these easily-connectorized, high-NA graded-index multimode fibers are ideal for short-distance (up to several kilometers) data communications applications such as local area networks. For graded-index multimode fiber used in data communications, the standard NAs are 0.20 for 50/125 μm fiber and 0.275 for 62.5/125 μm fiber.

critical *adj.* 临界的，关键的
performance *n.* 履行，执行，成绩，性能
specification *n.* 详述，规格，说明书，规范
particular *n.* 细节，详情；*adj.* 特殊的，特别的

vertical cavity surface emitting laser 垂直腔面发射激光器

alignment *n.* 队列
tolerance *n.* 公差

Defining NA

NA is a unitless quantity, as shown in Fig. 2. 7. It is derived from calculating the sine of the half angle (θ) of acceptance within the **cone** of light entering the fiber's core. Theoretical, NA may be expressed by the equation below:

$$NA = (n_1^2 - n_2^2)^{1/2}$$

where n_1 is the refractive index of the core and n_2 is the refractive index of the **cladding**. The refractive index of a material is defined as the ratio of the speed of light in a vacuum to the speed of light in that particular material.

cone n. ［数、物］锥形物，圆锥体

cladding n. 覆层

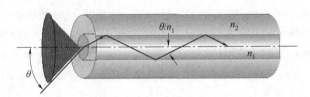

Fig. 2. 7 Numerical aperture defines the maximum angle（the "cone of acceptance"）at which light can be launched into a fiber

 Get to Work

Ⅰ. Matching

Directions：Match the words or expressions in the left column with the Chinese equivalents in the right column.

1. NA	A. 垂直腔面发射激光器
2. VCSELs	B. 数值孔径
3. LED	C. 激光
4. laser	D. 灯泡
5. lamp bulb	E. 发光二极管

Ⅱ. Understanding Checking

Directions：Mark Y（for YES）if the statement agrees with the information given in the text；N（for NO）if the statement contradicts the information given in the text.

（   ）1. Numerical aperture indicates the maximum angle at which a particular fiber can accept the light that will be transmitted through it.

（   ）2. Multimode fibers allow more modes of light to be transmitted, resulting in greater pulse spreading, or dispersion, and more bandwidth.

（   ）3. NA is a unitless quantity.

（   ）4. Theoretical NA may be expressed by the equation $NA = (n_1^2 - n_2^2)^{1/2}$, where n_1 is the refractive index of the core and n_2 is the refractive index of the cladding.

III. Passage Completion

Directions: Fill in each of the blanks with one of the words or expressions in the box, making changes if necessary.

viewpoint	radiant	range	nature
phenomenon	definition	fear	sure

Throughout human history, light has been something most of mankind has taken for granted. It is there throughout our lives for most of us, and (so we assume) will always be there in the familiar patterns we experienced as we grew up.

In the past, and in many countries even today, phenomena such as solar eclipses have been cause for great 1. _____, because they represent a break in that familiar pattern, cutting off the light from the sun for a while, and who could be 2. _____ if the sun would ever come back? Even in countries where an eclipse is an understood 3. _____, a solar eclipse is still an occasion for excitement and awe.

To gain any understanding of light itself, we need to step away from this mindset and examine light from a more scientific and objective 4. _____. Let's start with a dictionary of 5. _____ light, with some technical data included.

Light

Light is the form of 6. _____ energy that stimulates the organs of sight, having for normal human vision wavelengths 7. _____ from about 3,900 to 7,700 ångströms (Å) and traveling at a speed of about 186,300 mi/s.

1 Å = 10^{-8} cm (0.00000001 cm).

Of course, the above definition doesn't really tell us much. Before the speed of light and its wavelength in the electromagnetic spectrum were determined, the definition would have ended at the first comma, and that really would have told us nothing about the 8. _____ of light.

IV. Translation

Directions: Translate the following sentences from English into Chinese.

A new LED design employs a handy combination of light and phosphors to produce light whose color spectrum is not so different from that of sunlight.

Light emitting diodes convert electricity into light very efficiently, and are increasingly the preferred design for niche applications like traffic and automobile brake lights. To really make an impression in the lighting world, however, a device must be able to produce room light. And to do this one needs a softer, whiter, more color balanced illumination.

2. 4   Electromagnetic Radiation

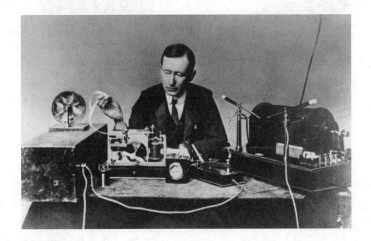

 Learning Objectives

In this section, you will:

1. understand the complete range of electromagnetic radiation.

2. find the reason why scientists said Marconi was wasting his time when he attempted to send radio waves from Cornwall to Newfoundland in 1902.

3. understand the reason for the spread and rebound of radio waves around the world.

4. indicate the height and timing of the different layers within the Heaviside layer.

5. understand that the presence of electrons and ions in the upper atmosphere makes it electrically conducting.

 Warm-up Activity

Directions: Every one of you is required to individually describe any picture shown in Fig. 2. 8 in one or two sentences, and then you are divided into groups of which each consists of five students or so. Any member of a group is responsible for describing one of the pictures one by one. Some groups will be invited to make a show in class.

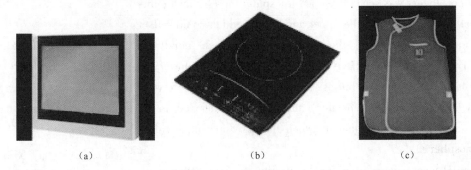

(a)   (b)   (c)

Fig. 2. 8   Pictures

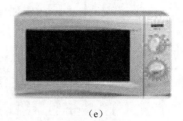

(d)   (e)

Fig. 2.8   Pictures（Continued）

 Text

The complete range of electromagnetic radiation is made up of **gamma rays**, X-rays, **ultraviolet rays**, ordinary visible light, **infrared**（heat）**radiation**, and radio waves. All these are given out by the sun, but some of them do not reach us, as the **atmosphere behaves** like a filter. It lets through visible light and certain radio waves; most of the other radiation is absorbed. This is fortunate for us, because gamma rays, X-rays, and ultraviolet rays can harm living things. Enough ultraviolet rays get through for us to detect them—they are responsible for our getting **sunburnt** when we expose ourselves too rapidly in sunbathing.

A well-known characteristic of light is that it travels in straight lines. The other electromagnetic radiations behave in a similar way. For this reason, scientists thought **Marconi** was wasting his time when he attempted to send radio waves from **Cornwall** to **Newfoundland** in 1902. It seemed obvious that the **curvature** of the earth across the **Atlantic** would prevent this.

However, the experiment was successful, and once again it had been shown that experiments should always be made to check **predictions** from theory.

A new theory was soon developed to explain the new facts; this was that a layer existed in the upper atmosphere which could reflect radio waves. In this way, they would be bounced round the world. This was called the **Heaviside** layer, named after one of the scientists who predicted it; but as later work showed that there were several layers at different heights, they are now referred to by letters, and the general region in which they are present is called the **ionosphere**.

The D-layer occurs at a height of fifty to sixty **miles** and

gamma ray 伽马射线

ultraviolet ray 紫外光射线

infrared *adj.* 红外线的；*n.* 红外线

radiation *n.* 发散，发光，发热，辐射

atmosphere *n.* 大气，空气

behave *v.* 举动，表现

sunburnt *adj.* 起晒斑的，晒黑的

Marconi 马可尼

Cornwall *n.* 康沃尔

Newfoundland 纽芬兰（岛）（加拿大的一个省）

curvature *n.* 曲率

Atlantic *n.* 大西洋

prediction *n.* 预言，预报

Heaviside 海维赛德

ionosphere *n.* 电离层

mile *n.* 英里（mi，英制长度单位，1 mi = 1.609344 km）

disappear during the night; the E-layer occurs at sixty to seventy miles and is weakened but still present at night. The F-layer extends from ninety miles upwards, and is present as two layers (F1 and F2) during the day. The layers are able to conduct electricity because the particles (atoms and molecules) of the atmospheric gases are **ionized**, that is to say, **split** into **ions** and electrons.

ionize *vt.* 使离子化; *vi.* 电离
split *v.* 分裂, 分离

An atom of any substance consists of a positively-charged central body (the nucleus) surrounded by negatively-charged electrons, like **planets** around a sun. Each electron bears a single negative charge; and usually the **nucleus** carries a number of positive charges equal to the number of electrons. Therefore, the whole atom is neither positive nor negative. If one or more electrons are removed, the remainder of the atom must be positively charged. It is possible to and extra electrons to the atom, and then we get negative ions.

ion *n.* 离子

planet *n.* [天] 行星
nucleus *n.* 核子 (nuclear 的复数)

Electric currents flowing in wires are streams of electrons running through the wire. Electrons also flow in a television tube, where they are made to hit the screen, causing a flash of light. Ions can also act as a current of electricity in liquids or gases, and the presence of electrons and ions in the upper atmosphere makes it electrically conducting. In fact, if there is a wind in the ionosphere, it will be an electric current.

 Get to Work

I. Matching

Directions: Match the words or expressions in the left column with the Chinese equivalents in the right column.

1. neutron	A. 原子
2. electron	B. 分子
3. ion	C. 离子
4. molecular	D. 中子
5. atom	E. 电子

II. Understanding Checking

Directions: Mark Y (for YES) if the statement agrees with the information given in the text; N (for NO) if the statement contradicts the information given in the text.

(　　) 1. The atmosphere lets through visible light and certain waves.

(　　) 2. The complete range of electromagnetic radiation is made up of gamma rays, X-rays, ultraviolet rays, ordinary visible light, infrared (heat) radiation, and radio waves.

( ) 3. The curvature of the earth across the Atlantic would prevent Marconi's attempt in 1902.

( ) 4. An atom of any substance consists of a positively-charged central body as well as negatively-charged electrons.

Ⅲ. Passage Completion

Directions: *Fill in each of the blanks with the corresponding expression in the box.*

Energy of the photon	The frequency of the wave
The index of refraction of the medium	Planck's constant
The velocity of light in open space	The wavelength of the wave

The basic transverse electromagnetic wave, involves both a varying electric field and a varying magnetic field, appearing at right angles to each other and to the direction of travel of the wave. Note especially that the electric and magnetic fields are not in phase with each other, but are rather 90° out of phase. Most books portray these two components of the total wave as being in phase with each other, but I find myself disagreeing with that interpretation, based on three fundamental laws of physics.

The total energy in the waveform must remain constant at all times. Any deviation from this condition constitutes a violation of this law.

The energy in a single photon is given by the following expression:

$$E = h\nu = \frac{hc}{n\lambda}$$

where:

$E = $ _____ ;

$h = $ _____ ;

$\nu = $ _____ ;

$c = $ _____ ;

$n = $ _____ ;

$\lambda = $ _____ .

Ⅳ. Para. Translation

Directions: *Translate the following paragraph from English into Chinese.*

Electric currents flowing in wires are streams of electrons running through the wire. Electrons also flow in a television tube, where they are made to hit the screen, causing a flash of light. Ions can also act as a current of electricity in liquids or gases, and the presence of electrons and ions in the upper atmosphere makes it electrically conducting. In fact, if there is a wind in the ionosphere, it will be an electric current.

Task Ⅲ　Skill Honing

2.5　Translation of Words and Phrases
词语的翻译

由于中英两种语言的内在结构不同，以及人们思维习惯、表达方式存在差异，因此在翻译过程中要考虑目的语读者的阅读习惯，如实地传达原文的思想内容，使译文通顺自然。在翻译词语时，通常有以下几种方法。

1. 直接翻译法。

保持原文的形式和内容，根据词语字面意思翻译出来。

Three types of printers are available：wire printers，ink jet printers and laser printers.

有三种类型的打印机：针式打印机、喷墨打印机和激光打印机。

2. 转换翻译法。

为了使译文更符合汉语的表达习惯，有时采用直译会觉得不自然，这就需要进行词类转换。

（1）转换为动词。

Attempts have been made to rename this machine.

人们一直在尝试给这台机器重新命名。

（名词 attempts 转换为动词"尝试"）

（2）转换为名词。

The average man who uses a laser printer could not explain how it *works*.

普通人只会使用激光打印机，却无法解释它的**工作原理**。

（动词 works 转换为名词"工作原理"）

（3）转换为副词。

Today's lasers are producing parts faster and with *higher quality*.

现代激光技术可以更快**更好地**制造各种部件。

（名词短语 higher quality 转换为副词"更好地"）

（4）转换为形容词。

The ozonosphere is of great *importance* to mankind.

臭氧层对人类是非常**重要的**。

（名词 importance 转换为形容词"重要的"）

3. 引申翻译法。

有时采用直译法无法表达词语内在包含的意义，这时就需要根据原文的意思，按照汉语的表达习惯，对词义进行引申。

This subject is still in its *infancy*.

这门学科仍处于**发展初期**。

（infancy 引申为"发展初期"）

4. 增词翻译法。

在译文中增加英语原文中没有的词，使表达的意思更准确更完整。

A drilled hole can be made accurate and smooth by a reamer.

钻好的孔可用绞刀加工，使其**尺寸**精确、**表面**光滑。

（增译"尺寸"和"表面"）

5. 省略翻译法。

在英语中有很多词，如冠词、介词、连词、代词，甚至动词等，在翻译时可以省略，这样使译文更自然流畅，避免生硬难懂。

A wave model ***is considered*** as a good model of the way in which light behaves.

波动模型是说明光的性能的一个很好的模型。

（省译 is considered）

6. 重复翻译法。

汉语注重形式，讲究文采，通常会重复某个词语加强语气、平衡结构，使语言富于表现力。

Fine-blanking press can be ***driven*** both mechanically and hydraulically.

精密冲切机既可以机械**传动**，也可液压**传动**。

（重复 driven）

Challenge

Directions：*Translate the following sentences into Chinese and pay special attention to the parts in bold.*

1. These are modern materials having greater uniformity ***and*** higher optical sensitivity.

2. It takes years to gain a ***mastery*** of the new technology.

3. The high intensity of a laser beam makes it a ***convenient*** drill.

4. The laser is shown ***briefly*** in the following chapter.

5. The ***micrometer*** is marked in hundredths of a millimeter.

Task Ⅳ   Job Task Challenges

2.6   Career Speaking：Meeting Guests

Ⅰ. Warm Up

Model

（A：Zhang Li   B：Mr. Mills）

A：Excuse me, are you Mr. Mills from New York?

B：Yes, I am.

A：Nice to meet you. I am Zhang Li, an *administrative* assistant of Wuhan Brightness Optoelectronics Corporation.

B：Nice to meet you, too, Miss Zhang. I really appreciate your coming to meet me.

A：You're welcome. How was your flight?

B：Great. It was *uneventful*.

A：I'm very glad to hear that. Shall I take you to the hotel now? Our car is waiting over there.

B：Yes, that would be wonderful!

A：We've reserved one suit for you at the *Shangri-La Hotel*, which is one of the best hotels in the city. I hope you have a good time there.

B：Thank you. You're really *considerate*.

A：My pleasure, Mr. Mills. In the evening, after the welcome party given by our manager, we'll take a car-ride round the city. At 8：30 tomorrow morning, I will drive to pick you up to our corporation. Is that all right?

B：All right. I'm here *at your disposal*.

Notes

administrative	*adj.* 行政的，管理的
uneventful	*adj.* 太平无事的
Shangri-La Hotel	香格里拉饭店
considerate	*adj.* 考虑周到的
at your disposal	任你处理；听你安排

Ⅱ. Matching

Directions：Please choose the corresponding sentences from the box to fill in the blanks. Then, play roles with your partner.

（A：Wang Yun   B：Zhang Wen）

A：_____

B：Yes, I am.

A：Hello! _____ Our manager has asked me to

come and meet you.

B：_____

A：You're welcome. _____

B：Thank you.

A：Not at all. _____

B：Yes，very pleasant.

A：The waiting room is over there. Let's take a short rest. _____

S1：Let me help you with your baggage.

S2：Did you have a good trip?

S3：Excuse me，are you Miss Zhang Wen from Sichuan?

S4：And then I'll drive you to the hotel.

S5：Nice to meet you. I'm Wang Yun，the secretary of Yichang Foreign Trade Company.

S6：Nice to meet you，too，Miss Wang. Thank you for meeting me.

Ⅲ. Closed Conversation

Directions：*Please fill in the blanks according to the Chinese. Then，play roles with your partners.*

（A：Lily　　B：Ben Green　　C：Jerry）

A：Good morning，Mr. Green. _____ （请让我向你介绍我的同事杰利，他也是来接你的。）

B：How do you do，Jerry?

C：_____ （你好，格林先生，欢迎来到武汉。）

B：Thank you. I prefer Ben，if that's OK.

C：Ben，how was the flight?

B：_____ （不太好。由于糟糕的天气航班延误了好几个小时。）

C：After a long flight，you need a good rest.

A：Ben，if you don't mind，while you and Jerry get acquainted，_____ _____ （我把车开过来，请稍等。）

B：All right. Jerry，when shall we get down to business?

C：_____ （我已经准备了日程表，请过目，好吗?）

B：OK. Let me see. Tomorrow afternoon we will have a meeting，right?

C：Yes. _____ （会议定于明天下午两点举行，我会在下午一点半来接您。）

B：OK. _____ （从日程表上看，我发现你们已经安排了两天的时间带我去市里各处游览。谢谢，你们真是想得太周到了。）

A：_____ （别客气。我们希望您在武汉过得愉快。车来了，我们走吧。）

IV. Semi-open Conversation

Directions: *Please fill in the blanks with your own words. Then, play roles with your partners.*

(A: Susan　　B: Jane)

A: Hi, Jane. _____

B: Me too. _____

A: My pleasure. _____

B: It was a pleasant trip but I'm a little tired.

A: _____ Then, you'll have a good rest.

B: Thank you. You're very thoughtful.

A: Don't mention it. _____ Let's go.

V. Live Show

Challenge 1

Directions: *Mr. Zhang is asked to meet Mr. Smith from San Francisco at the airport. They greet each other and then Mr. Zhang drives Mr. Smith to the hotel. Please act it out with your partner or partners. The model and the useful expressions below are just for your reference.*

Challenge 2

Directions: *David, together with his colleague Miss Wu, on behalf of the manager of Happiness Import & Export Company, is asked to pick up John. David greets his old friend John and introduces Miss Wu to him. They three talk about the flight and the arrangement of business. Please act it out with your partner or partners. The model and the useful expressions below are just for your reference.*

Model

(A: Henry Jones　　B: Sally　　C: Liu Xin)

B: Good morning, Henry. Delighted to see you again.

A: Me too, Sally. It's been quite a while, hasn't it?

B: Right. It's been almost one year since I saw you last time but you haven't changed a bit.

A: Thank you. Neither have you.

B: I would like to introduce Mr. Liu Xin to you. Mr. Liu is the manager of our *import division*. Mr. Liu, this is Henry Jones from Good Luck Import and Export Co. Ltd.

C: Glad to meet you, Mr. Jones. We've been looking forward to seeing you.

A: Glad to meet you, too. Mr. Liu. Thank you for your coming to meet me.

C: It's my pleasure. Did you enjoy your trip?

A: Yes, it was a pleasant trip. But I am feeling a little *jet lag*.

C: If you take a rest, I'm sure you'll be fine tomorrow, Mr. Jones.

A: Henry, please. "Mr. Jones" makes me sound ancient.

C: OK, Henry. You can also call me Peter, my English name.

B: Henry, please give me your baggage *check*. I'll go and *claim the baggage* for you.

A: Here you are. Thanks.

B: You're welcome. I'll be back soon.

C：How do you like the weather here, Henry?

A：It's a lovely day. The weather here is quite similar to that in our city.

C：As far as I know, you have been in Wuhan for several times. Do you feel accustomed?

A：Yes. I think it's all right. I like Wuhan. It's a *fantastic* place.

C：You must have visited many *scenic spots and historical sites*.

A：Frankly speaking, I am always too busy to spare any time for a tour.

C：If you like, we'll set up a *sightseeing* program for you.

A：That's great. It's very kind of you, Peter.

C：Please think nothing of it. Sally will take care of you during your stay in Wuhan. Don't hesitate to call her if you need any help. I hope you *feel at home* here.

A：Thank you very much. Sally is my good friend. She is really warm-hearted and thoughtful.

C：I'm glad to hear that. Oh, here she comes. We have a car waiting outside to take you to your hotel. Shall we go now?

A：Yes, let's go.

Notes

import division	进口部
jet lag	时差反应
check	*n.* 用以识别的票证或纸片
claim the baggage	领行李
fantastic	*adj.* 极好的
scenic spots and historical sites	名胜古迹
sightseeing	*n.* 观光
feel at home	像在家一样感到轻松舒适

Useful Expressions

1. 见面寒暄。

Hi, how's your business?

Delighted to see you again.

It's a pleasure to meet you at last.

I've heard so much about you.

It's very kind of you to come here to meet me.

2. 相互介绍。

Allow me to introduce myself.

May I introduce Tom to you?

I'd like you to meet my colleague Mary.

This is Linda from England.

3. 旅途情况。

Did you have a good trip?

How was your flight?

Yes, it was a very smooth flight.

Not so good. There was a thunderstorm so the flight was delayed for several hours.

This one was uneventful, except for a little turbulence here and there.

4. 去宾馆的路上。

I hope you'll enjoy your stay here.

I've made a reservation at the hotel you used last time.

I've booked a room for you at the Hilton Hotel. Single, for 9 days.

I'll have the car brought here, so please wait a moment.

5. 日程安排。

It's just the matter of the schedule, that is, if it is convenient for you right now.

We'll leave some evenings free, if it is all right with you.

The meeting is scheduled for 3：30 this afternoon. I will pick you up at 3：00 p.m.

I have arranged your schedule. I hope it's suitable.

We've arranged one afternoon in the time schedule to show you around the city.

2.7　Writing：Letter of Thanks

在人际交往中，对别人的帮助、鼓励、馈赠的礼物、提供的服务等表示真诚的感谢是很有必要的，既能体现自身的道德素质，又能拉近人与人之间的关系。除了当面道谢和电话道谢外，用别致的卡片或短信也能表达自己的心意。在写感谢信时要特别注意以下几点。

1. 开门见山，言简意赅，说明感谢的原因。
2. 措辞自然，态度诚恳，不可夸大其词、不切实际。
3. 表达自己的良好祝愿。

Sample 1

May 22, 2011

Dear Mr. Smith,

I am writing this letter to thank you for the wonderful time I spent during my stay in Wuhan. I was **overwhelmed** by your warm hospitality which made me feel at home, and I was impressed by the visit to your company which was very informative and **inspiring**.

I would like to express my best wishes for the success in your career.

Yours sincerely,
Bill Lawrence

Notes

overwhelm	*v.* 压倒
inspiring	*adj.* 令人鼓舞的，振奋人心的

Sample 2

Dec. 19, 2011

Dear Miss Wang,

Thank you for your letter of Dec. 16, enclosing the catalogues and price lists, which fascinate us very much. I find the products with new designs and favorable prices gain an edge over your competitors. We really appreciate your sincerity and look forward to establishing trade relations with your company.

Best wishes.

Yours faithfully,
Nancy Smith

Useful Expressions

1. Many thanks for your...

2. Please accept my sincere appreciation for...

3. I am writing this letter to thank you for...

4. Thank you more than I can say.

5. I take this opportunity to express to you my deep appreciation for...

Challenge

Directions: *Your colleague Miss Zhang sent you a gift for your wedding. You write a letter to express your thanks.*

Task Ⅴ　Output and Evaluation

Electromagnetic Radiation					
Target	Understand and express professional knowledge				
Requirement	Make an English presentation according to lesson 2. 4. Draw a mind map. Produce micro-video to express professional knowledge. Add accurate Chinese and English subtitles to the video.				
Contents	Electromagnetic Radiation： Complete range of electromagnetic radiation Experiment Ionosphere Conducting				
Group： _____	**Project**	**Name**	**Software**	**Score**	**Requirements**
	Mind map				The logic should be sound, and the keywords used should be accurate.
	Script				The format should adhere to the specified guidelines, and it should cover all necessary contents.
	PPT				It should be consistent with the logic presented in the mind map. Clear and concise contents/Consistent formatting/Limited text/Engaging graphics and animations.
	Speaking				Clear/fluent
	Subtitle				They should be bilingual, correct and synchronized with the student's speech.
	MP4				The video should effectively convey knowledge, be accurate, and visually engaging.
Operating environment	Win7/Win8/Win9/Win10/Win11/Mobile phone				
Product features					

Electromagnetic Radiation
思维导图

Electromagnetic Radiation
微课

Script reference

Electromagnetic Radiation

标题	专业内容（讲稿与中英字幕）	PPT 中需显示的文字	表现形式	素材
每个知识点的要点，或者小节名字	做报告时候的英文讲稿 视频中的英文字幕	需要特别提醒的文字	讲解到某个地方或某句话时，出现的图片、动画或者文字	照片，或者视频
Contents	**Electromagnetic Radiation** Contents： 1. Complete Range of Electromagnetic Radiation 2. Experiment 3. Ionosphere 4. Conducting **电磁辐射** 内容： 1. 电磁波谱 2. 实验 3. 电离层 4. 导电	**Electromagnetic Radiation** 1. Complete Range of Electromagnetic Radiation 2. Experiment 3. Ionosphere 4. Conducting	文字	
1. Complete range of electro magnetic radiation	The complete range of electromagnetic radiation is made up of gamma rays, X-rays, ultra-violet-rays, ordinary visible light, infra-red (heat) radiation, and radio waves. 整个电磁辐射范围由 γ 射线，X 射线，紫外线，普通可见光线，红外辐射和无线电波组成。	Complete range of electromagnetic radiation： gamma rays, X-rays, ultra-violet-rays, ordinary visible light, infra-red (heat) radiation, and radio waves	图示	图 1　电磁波谱图
	All these are given out by the sun, but some of them do not reach us, as the atmosphere behaves like a filter. 所有这些射线都由太阳发射出来，但是其中一些射线并不能到达我们这里，因为大气层起到了过滤作用。		图示	图 2　大气过滤
2. Experiment 实验	A well-known characteristic of light is that it travels in straight lines. The other electromagnetic radiations behave in a similar way. 光的一个尽人皆知的特性是直线传播。其他电磁辐射也一样。	Experiment	文字	

续表

标题	专业内容（讲稿与中英字幕）	PPT 中需显示的文字	表现形式	素材
2. Experiment 实验	For this reason, scientists thought Marconi was wasting his time when he attempted to send radio waves from Cornwall to Newfoundland in 1902. 因此，科学家们当时认为，马可尼在 1902 年想把线电波由康沃尔发送到纽芬兰是在浪费时间。	Marconi He attempted to send radio waves from Cornwall to Newfoundland in 1902.	图示	图3　科学家马可尼
	It seemed obvious that the curvature of the earth across the Atlantic would prevent this. 横跨大西洋的地球弧度会使这个想法不能实现，这似乎是显而易见的。	It seemed obvious that the curvature of the earth across the Atlantic would prevent this.		图4　地球曲率
	However, the experiment was successful, and once again it had been shown that experiments should always be made to check predictions from theory. 然而实验成功了，它又一次说明，从理论上推出的预见是需要通过实验来检验的。	However, the experiment was successful.		
3. Ionosphere 电离层	A new theory was soon developed to explain the new facts; This was that a layer existed in the upper atmosphere which could reflect radio waves. In this way, they would be bounced round the world. 为了解释这种新的事实，很快又提出了一种新理论，即在大气层的上部，有一个能反射无线电波的层次。这样，无线电波可以绕地球多次反射。	A new theory was soon developed to explain the new facts; this was that a layer existed in the upper atmosphere which could reflect radio waves. In this way, they would be bounced round the world.		图5　无线电波多次反射图
	This was called the Heaviside layer, after one of the scientists who predicted it; But as later work showed that there were several layers at different heights, they are now referred to by letters, and the general region in which they are present is called the ionosphere. 这一层叫海维赛层，是以一位对此作出预言的科学家的名字命名的。但是由于后来的研究表明，在不同的高度分别有好多层。现在，各层都用字母来表示，而它们所处的总区域叫作电离层。	There were several layers at different heights, they are now referred to by letters, and the general region in which they are present is called the ionosphere.		图6　电离层

续表

标题	专业内容（讲稿与中英字幕）	PPT 中需显示的文字	表现形式	素材
D-layer	The D-layer occurs at a height of fifty to sixty miles and disappear during the night. D 电离层位于 50～60 英里的高度，夜间便消失了。	The D-layer occurs at a height of fifty to sixty miles and disappear during the light.	图示 强调 D 层	图 6　电离层
E-layer	the E-layer occurs at sixty to seventy miles and is weakened but still present at night. E 电离层位于 60～70 英里的高度，夜间减弱，但一直还存在。	The E-layer occurs at sixty to seventy miles and is weakened but still present at night.	图示 强调 E 层	图 6　电离层
F-layer	The F-layer extends from ninety miles upwards, and is present as two layers（F1 and F2）during the day. F 电离层从 90 英里向上延伸，白昼是两层（F1 和 F2）。	The F-layer extends from ninety miles upwards, and is present as two layers（F1 and F2）during the day.	图示 强调 F 层	图 6　电离层
4. The layers are able to conduct electricity	The layers are able to conduct electricity because the particles（atoms and molecules）of the atmospheric gases are ionized, that is to say, split into ions and electrons. 各层均能导电，因为大气粒子（原子和分子）被电离，也就是说被分裂为离子和电子了。	The layers are able to conduct electricity because the particles（atoms and molecules）of the atmospheric gases are ionized, that is to say, split into ions and electrons.		图 7　原子模型 
	The presence of electrons and ions in the upper atmosphere makes it electrically conducting. In fact, if there is a wind in the ionosphere, it will be an electric current. 在大气上层存在电子和离子，就使其具有导电性。事实上，如果电离层中有风的话，那大气就是电流。			图 7　原子模型

Task VI　Knowledge Expansion

2.8　What Is Light

Throughout human history, light has been something most of mankind has taken for granted. It is there throughout our lives for most of us, and (so we assume) will always be there in the familiar patterns we experienced as we grew up.

In the past, and in many countries even today, phenomena such as solar eclipses have been causing for great fear, because they represent a break in that familiar pattern, cutting off the light from the sun for awhile, and who could be sure if the sun would ever come back? Even in countries where an eclipse is an understood phenomenon, a solar eclipse is still an occasion for excitement and awe.

To gain any understanding of light itself, we need to step away from this mindset and examine light from a more scientific and objective viewpoint. Let's start with a dictionary definition of light, with some technical data included.

Light

Light is the form of radiant energy that stimulates the organs of sight, having for normal human vision wavelengths ranging from about 3,900 to 7,700 ångströms (Å) and traveling at a speed of about 186,300 mi/s.

$1 \text{ Å} = 10^{-8}$ cm (0.00000001 cm).

Of course, the above definition doesn't really tell us much. Before the speed of light and its wavelength in the electromagnetic spectrum were determined, the definition would have ended at the first comma, and that really would have told us nothing about the nature of light.

So, rather than look at more definitions, let's move on to explore some of the basic properties of light as we know them today, so we can better understand not only how light will behave, but also something of why it behaves as it does. This will give us a chance to predict how light may behave under various circumstances and conditions.

Exercise

Directions: Answer the following question.

Can you tell us what is light in your own words?

2.9　Convex Lens

The most commonly-seen type of lens is the convex lens. This type of lens is often used for close examination of small objects, such as rare stamps or coins. Children often use such a lens to concentrate sunlight to burn small pinholes in pieces of paper. That result by itself shows the power of concentrated light from the sun. But there must be more to it than that. Let's see if we can define the behavior of lenses a bit more specifically.

Fig. 2. 9 shows a double convex lens with several rays of light approaching from its left. We show each ray as a different color here, simply to more easily follow each ray's progress. We will assume that the lens is made of glass with a nominal index of refraction of 1. 50.

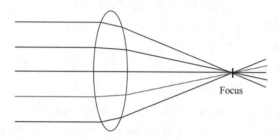

Fig. 2. 9　A double convex lens

The rays are parallel as they approach the lens. As each ray reaches the glass surface, it refracts according to the effective angle of incidence at that point of the lens. (See the pages on refraction for the definitions and descriptions of these terms.) Since the surface is curved, different rays of light will refract to different degrees; the outermost rays will refract the most.

As the light rays exit the glass, they once again encounter a curved surface, and refract again. This further bends the rays of light towards the centerline of the lens (which coincides with the green light ray in the figure).

Exercises

Directions: *Translate the following words or phrases into Chinese.*

1. convex lens

2. nominal index of refraction

3. effective angle

4. degree

5. ray

Unit 3　Laser

In this unit, you are going to learn the following contents:

Task Ⅰ Technical Principles and Equipment Cognition

3. 1　This Is Laser

3. 2　Laser Construction

Task Ⅱ　Technology Application Understanding

3. 3　Applications of Lasers

3. 4　Internet and Lasers

Task Ⅲ　Skill Honing

3. 5　Translation of Noun Clauses

Task Ⅳ　Job Task Challenges

3. 6　Career Speaking: Introducing Enterprises

3. 7　Writing: Letter of Congratulation

Task Ⅴ　Output and Evaluation

Task Ⅵ　Knowledge Expansion

3. 8　A Prospectus about Laser Machines

3. 9　Using Lasers

Unit 3

Laser

Task I   Technical Principles and Equipment Cognition

3.1   This Is Laser

Learning Objectives

In this module, you will:

1. understand the meaning of the word "laser".

2. learn about applications of laser.

3. master differences between laser and other light.

Warm-up Activity

Directions：Please make up a short dialogue or story according to Fig. 3. 1.

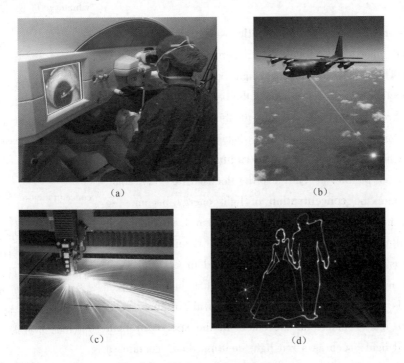

(a)　　　　　　　　　　　　(b)

(c)　　　　　　　　　　　　(d)

Fig. 3. 1　Dialogue/story pictures
(a) Laser surgery；(b) Laser weapon；(c) Laser cutting；(d) Laser dancing

Text

Most people know the word "laser", but do they know what it really is? What's the difference between ordinary light and laser and what does laser really **stand for**? Let's start with the last question.

Laser is an **acronym**, which is a word made up of **initial** letters. You could use the complete name：light **amplification** by **stimulated emission** of **radiation**, but that's a bit **awkward**. Let's **keep to** laser.

Almost everyone probably knows that the police use laser when they measure speed. At least, many drivers that have **exceed**ed the **speed limit** know about it, but how many know that you also use laser several times during an ordinary day? You'll find it in **CD** players, laser printers and much, much more.

You often find laser in action movies where the hero has to escape the laser beams when he's trying to solve a **thrilling**

stand for 代表，象征
acronym *n.* 首字母缩略词

initial *adj.* 最初的，词首的
amplification *n.* 扩大
stimulate *vt.* 刺激，激励
emission *n.* 散发，发射
radiation *n.* 发散，发光，辐射
awkward *adj.* 难使用的，笨拙的
keep to 坚持，信守
exceed *vt.* 超越
speed limit 限速
CD（compact disk）光盘
thrilling *adj.* 毛骨悚然的，颤动的

problem. The power contained in laser is both **fascinating** and frightening.

Now, let's turn to the difference between laser light and other light.

Light is really an **electromagnetic** wave. Each wave has brightness and color, and **vibrates** at a certain angle, so-called **polarization**. This is also true for laser light but it is more **parallel** than any other light source. Every part of the beam has (almost) the exact same direction and the beam will therefore **diverge** very little. With a good laser an object at a distance of 1 km can be **illuminated** with a dot about 60 **mm** in **radius**.

As it is so parallel it can also be focused to very small **diameters** where the **concentration** of light energy becomes so great that you can cut, **drill** or turn with the beam. It also makes it possible to illuminate and examine very tiny details. It is this property that is used in **surgical appliances** and in CD players. It can also be made very **monochromic**, so that just one light **wavelength** is present. This is not the case with ordinary light sources. White light contains all the colors in the **spectrum**, but even a colored light, such as a red light emitting diode contains a continuous **interval** of red wavelengths.

On the other hand, laser emissions are not usually very strong when it comes to energy content. A very powerful laser of the kind that is used in a laser show does not **give off** more light than an ordinary **streetlight**; the difference is in how parallel it is.

fascinating *adj.* 迷人的，着魔的

electromagnetic *adj.* 电磁的
vibrate *v.* （使）振动，（使）摇摆
polarization *n.* ［物］偏振（现象）；分化
parallel *adj. & n. & v.* 平行
diverge *vi.* 分叉，分歧
illuminate *vt.* 照明，说明
mm *abbr.* millimeter 毫米
radius *n.* 半径，范围，辐射光线
diameter *n.* 直径
concentration *n.* 集中，专心，浓缩，浓度
drill *v.* 训练，钻孔
surgical *adj.* 外科的，手术上的
appliance *n.* 用具，器具
monochromic *adj.* 单色的
wavelength *n.* ［物］波长
spectrum *n.* 光谱，型谱，频谱

interval *n.* 间隔，距离，幕间休息

give off *v.* 发出（蒸汽、光等）
streetlight *n.* 街灯

 Get to Work

Ⅰ. **Matching**

Directions: Match the words or expressions in the left column with the Chinese equivalents in the right column.

1. air-proof   A. 固定光束
2. beam axis   B. 激光探测器
3. optical guiding system   C. 远红外区
4. fixed beam   D. 光学头
5. perfect lens   E. 密封的
6. laser detector   F. 玻璃棒
7. optical path   G. 射束轴
8. optical heads   H. 理想透镜

9. far-infrared region   I. 光程
10. glass rod   J. 光制导系统

Ⅱ. Understanding Checking

Directions：*mark Y（for YES）if the statement agrees with the information given in the text*；*N（for NO）if the statement contradicts the information given in the text.*

（   ）1. Since "light amplification by stimulated emission of radiation" seems to go beyond our understanding，let's give it up.

（   ）2. Laser finds its application in measuring car speed，CD players，printers and so on.

（   ）3. White light contains all the colors in the spectrum.

（   ）4. Compared with other light，laser diverges little.

（   ）5. Light given off by laser is more than that of an ordinary streetlight.

Ⅲ. Passage Completion

Directions：*Fill in each of the blanks with one of the words or expressions in the box，making changes if necessary.*

introduce	develop	do	categorize	tool
trend	supplier	settle	overview	apply

The TRUMPF Group is one of the world's leading corporations in production technology，with a total of approximately 6,500 employees at 43 locations in 23 countries. Our innovations set the 1. _____ ：we are the leader for industrial laser technology and laser systems in the global market，and are among the largest machine tool 2. _____ worldwide，because we never have 3. _____ for second-rate solutions. This 4. _____ not only to our products，but also to the names we give them.

We 5. _____ the names of our machines，lasers，programming systems，tools and accessories parallel with our product range in the past. As our product range grew over the years，so 6. _____ the number of product names.

For that reason，we 7. _____ a new name concept now. This will initially apply to our two largest business fields：machine 8. _____ and laser technology.

These new names are intended to provide an improved 9. _____ of our product program structure. Machines can now 10. _____ easily according to their new distinctive names.

Ⅳ. Phrase Translation

Directions：*Complete the sentences by translating into English according to the Chinese given in brackets.*

1. Before long the laser's distinctive qualities— _____ （能生成高强度、单一波长的窄光束）— were recognized.

2. Lasers are everywhere：from research laboratories at the cutting edge of quantum physics to _____ （诊所、超市收银处和电话网络）.

3. He analyzed the spectrum of light emitted and found _____
（其频率范围明显缩小）.

4. Theodore Maiman _____（使第一台激光成功运行）on
May 16, 1960 at the Hughes Research Laboratory in California.

5. The scientist did not report having seen a bright beam of light, _____
_____（这是人们所预期的激光器特征）.

V. Sentence Remaking

Directions: *Simulate the following sentence patterns according to an example provided from the text.*

1. … is …, which is … made up …

E. g. Laser is an acronym, which is a word made up of initial letters.

_____ is/are _____, which is _____made up of _____.

2. …find … in … where …

E. g. You often find laser in action movies where the hero has to escape the laser beams.

_____ find/finds _____ in _____ where _____.

3. … make it possible to …

E. g. It also makes it possible to illuminate and examine very tiny details.

_____ makes/make it possible to _____.

4. It is … that …

E. g. It is this property that is used in surgical appliances and in CD players.

It is _____ that _____.

5. On the other hand, … are/is … when it comes to …

E. g. On the other hand, laser emissions are not usually very strong when it comes to energy content.

On the other hand, _____ are/is useful when it comes to _____.

VI. Para. Translation

In 1957, Charles Hard Townes and Arthur Leonard Schawlow, then at Bell Labs, began a serious study of the **infrared maser**（红外线微波激射器）. As ideas were developed, **frequencies**（频率）were abandoned with focus on visible light instead. The concept was originally known as an "**optical maser**"（光量子放大器,）. Bell Labs filed a **patent application**（专利申请）for their proposed optical maser a year later. Schawlow and Townes sent a manuscript of their theoretical calculations to *Physical Review*, which published their paper that year.

VII. Mini Imitative Writing

Directions: *The following paragraph originates from the text. Please imitate it and write another one. The following topics may be just for reference*: MASER, LED, VCD, CCD, MOS, *etc.*

Laser is an acronym, which is a word made up of initial letters. You could use the complete name: light amplification by stimulated emission of radiation, but that's a bit awkward. Let's keep to laser.

3. 2　Laser Construction

Laser Construction

动画

 Learning Objectives

In this module, you will:

1. describe the basic structure of a laser.

2. understand the function of pump source, examples of pump sources.

3. indicate pump sources of helium-neon (HeNe) lasers, Nd：YAG lasers, excimer lasers.

4. understand the role and principle of the gain medium, examples of different kinds of gain medium.

5. describe the configuration of the optical resonator (simple / complex).

6. analyze the working process of the optical resonator.

 Warm-up Activity

Directions：Please write an individual signature（个性签名）for each of the following machines as if they were personified. You may discuss with your partner with help of the example in Fig. 3. 2（a）. Some of you will be invited to show your created individual signatures to the whole class.

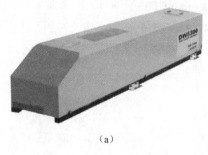

（a）

（b）

Individual signature：My predecessor was a top one in the working guys, so are his offspring.

Individual signature：_____

_____

_____.

Fig. 3. 2　Signature for machines

（a）Solid-state laser；（b）Gas laser

(c)

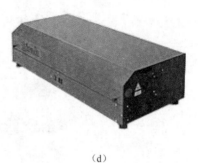

(d)

Individual signature：_____

_____

_____．

Individual signature：_____

_____

_____．

Fig. 3.2　Signature for machines（Continued）

（c）Excimer laser；（d）Dye laser

Text

　　Laser is constructed from three principal parts：An energy source（usually referred to as the pump or pump source，as shown in Fig. 3.3），a **gain medium** or laser medium，and two or more mirrors that form an **optical resonator**.

Pump Source

　　The pump source is the part that provides energy to the laser system. Examples of pump sources include electrical **discharges**, **flashlamps**，**arc lamps**，light from another laser，chemical reactions and even explosive devices. The type of pump source used principally depends on the gain medium，and this also determines how the energy is transmitted to the medium. A **helium-neon** laser uses an electrical discharge in the helium-neon gas mixture，a **Nd：YAG** laser uses either light focused from a **xenon flash lamp** or **diode lasers**，and **excimer** lasers use a chemical reaction.

Gain Medium

Fig. 3.3　Pump source

gain medium 增益介质

resonator *n.* 谐振腔

optical resonator 光学谐振腔

discharge *n.* 放电

flashlamp *n.* 闪光灯

arc lamp 弧光灯

helium *n.* 氦（化学符号为 He）

neon *n.* 氖（化学符号为 Ne）

xenon *n.* 氙

Nd：YAG（neodymium-doped yttrium aluminum garnet）掺钕钇铝石榴石

xenon flashlamp 氙气闪光灯

diode laser 二极管激光器

excimer *n.* 受激准分子

The gain medium is the major determining factor of the wavelength of operation, and other properties of the laser. The gain medium is excited by the pump source to produce a **population inversion**, and it is in the gain medium that **spontaneous** and **stimulated emission** of photons takes place, leading to the phenomenon of optical gain, or amplification.

population inversion 粒子数反转
spontaneous *adj.* 自发的, 自然产生的
stimulated emission 受激发射

Examples of different gain media include liquids, gases, and solids.

Liquids, such as **dye**. The exact chemical configuration of the dye molecules determines the operation wavelength of the dye laser.

dye *n.* 染料; *vt.* 染

Gases, such as **carbon dioxide** and mixtures such as helium-neon. These lasers are often **pumped** by electrical discharge.

Solids, such as **crystals** and glasses. The solid host materials are usually **doped** with an **impurity**.

carbon dioxide *n.* 二氧化碳
pump *n.* 泵; *vt.* 抽吸, 抽运
crystal *n.* 水晶
dope *v.* 加入掺杂剂
impurity *n.* 杂质, 混杂物, 不纯

Semiconductors, a type of solid, in which the movement of electrons between material with differing **dopant** levels can cause laser action.

dopant *n.* 掺杂物, 掺杂剂

Optical Resonator

The optical resonator (Fig. 3.4), or optical **cavity**, in its simplest form is two parallel mirrors placed around the gain medium which provide **feedback** of the light. The mirrors are given optical **coatings** which determine their reflective properties. Typically, one will be a **high reflector**, and the other will be a **partial reflector**. The latter is called the output **coupler**, because it allows some of the light to leave the cavity to produce the laser's output beam.

cavity *n.* 空穴, 腔

feedback *n.* 反馈, 反应
coating *n.* 涂料
reflector *n.* 反射体, 反射镜
high reflector 全反射镜
partial reflector 半透反射镜
coupler *n.* 耦合器

Fig. 3.4　Optical resonator

Light from the medium, produced by **spontaneous emission**, is reflected by the mirrors back into the medium, where it may be amplified by stimulated emission. The light may reflect from the mirrors and thus pass through the gain medium many hundreds of times before exiting the cavity. In more complex lasers, configurations with four or more mirrors forming the cavity are used. The design and alignment of the mirrors **with respect to** the medium is crucial to determining the exact operating wavelength and other **attributes** of the laser system.

spontaneous emission 自发发射

with respect to 关于
attribute *n.* 属性，特征

 Get to Work

Ⅰ. Matching

Directions: *Match the words or expressions in the left column with the Chinese equivalents in the right column.*

1. antielectron
2. basic frequency
3. fiber sensor
4. optical filter
5. photo-electron
6. diode pumping
7. input amplifier
8. collimated beam
9. acousto-optic cavity
10. alternating current

A. 声光腔
B. 二极管抽运
C. 反电子
D. 准直光束
E. 交流电
F. 光电子
G. 主频率
H. 光纤感应器
I. 输入放大器
J. 滤光器

Ⅱ. Understanding Checking

Directions: *Choose the best answer for each statement below according to the text.*

1. Laser is made up of the following parts: a pump source, a gain medium, and _____ .

A. the pump   B. an energy source   C. laser medium   D. an optical resonator

2. Pump sources include arc lamps, light from another laser, electrical discharges, and even _____ .

A. xenon   B. explosive devices   C. a coupler   D. a high reflector

3. The gain medium is _____ by the pump source to produce a population inversion.

A. stimulated   B. motivated   C. encouraged   D. inspired

4. Different gain medium contain liquids, gases, _____ , and semiconductors.

A. dopants   B. crystal   C. solids   D. coating

5. Optical coatings _____ the reflective properties of the mirrors in the optical resonator.

A. depend on   B. determine   C. allow for   D. take advantage of

III. Passage Completion

Directions: *Fill in each of the blanks with one of the words or expressions in the box, making changes if necessary.*

measurement	pulse	generate	prove	considerable
resonator	set up	lamp	principle	operate

On May 16, 1960, the laser was born. Laser light was 1. _____ for the first time. The experiment 2. _____ by Theodore H. Maiman was astonishingly simple with a short, thick ruby rod as an active media. The end surfaces of the rod were used as the 3. _____. A coiled flash 4. _____ was used for excitation. The first laser developed by Maiman was a 5. _____ solid-state laser.

Maiman's ruby laser could only 6. _____ with minimal pulse repetitions. But the pulse power and energy were already quite 7. _____: A pulse penetrated through a packet of razor blades. This was no mere game. The number of penetrated razor blades was the relative unit of measurement for the pulse energy because there were no 8. _____ devices at the time.

Its successors 9. _____ themselves particularly in production technology. Tens of thousands of them do their work around the world on a higher technical level but still based on the same 10. _____.

IV. Para. Translation

Directions: *Complete the sentences by translating into English according to the Chinese given in brackets.*

1. Some optical devices may be placed within the optical resonator to _____ (产生多重效应) on the laser output.

2. Some so-called lasers rely on very high optical gain without _____ (需要光反射折回增益介质).

3. Since they do not use optical feedback, these devices _____ (不归类为激光器).

4. Semiconductor lasers are typically very small, and can _____ (抽运只需普通电流).

5. In 1947, W. E. Lamb and R. C. Retherford _____ (首次展示了受激发射).

V. Sentence Remaking

Directions: *Simulate the following sentence patterns according to an example provided from the text.*

1. ... is/are constructed from ...

E. g. Laser is constructed from three principal parts.

_____ is/are constructed from _____.

2. This determines how …

E. g. This determines how the energy is transmitted to the medium.

This determines how _____.

3. … is/are … by … to …

E. g. The gain medium is excited by the pump source to produce a population inversion.

_____ is/are _____ by _____ to _____.

4. … takes/take place, leading to …

E. g. Spontaneous and stimulated emission of photons takes place, leading to the phenomenon of optical gain.

_____ takes/take place, leading to _____.

5. … is crucial to …

E. g. The design and alignment of the mirrors is crucial to determining the exact operating wavelength and other attributes of the laser system.

_____ is crucial to _____.

VI. Para. Translation

Directions: *Translate the following paragraph from English into Chinese.*

Most solid state lasers are rod lasers (laser rods), i. e. , lasers with a rod-shaped doped laser crystal as the gain medium. Strictly speaking, a rod always has a cylindrical (圆柱的) shape, but the term is sometimes also used for crystals with rectangular shapes. Its ends are normally either perpendicular (垂直的) to the beam axis, or are Brewster-angled (布鲁斯特角) for suppressing (抑制) parasitic reflections and ensuring a stable linear polarization. Rod lasers can be either end-pumped or side-pumped. Cooling is the simplest for end pumping, where the outer surface can be fully surrounded by a water-cooled heat sink.

VII. Mini Imitative Writing

Directions: *The following paragraph originates from the text. Please imitate it and write another one.*

The pump source is the part that provides energy to the laser system. Examples of pump sources include electrical discharges, flashlamps, arc lamps, light from another laser, chemical reactions and even explosive devices. The type of pump source used principally depends on the gain medium, and this also determines how the energy is transmitted to the medium.

Task Ⅱ　Technology Application Understanding

3.3　Applications of Lasers

Learning Objectives

In this module, you will:

1. understand industrial applications of lasers.

2. master laser application in environment and communications.

3. learn about laser's role in research.

Warm-up Activity

Directions: Describe the pictures in Fig. 3.5. Use the provided phrases, whose meaning is equivalent to "应用于", to orally express "激光应用于……领域" in English to your partner with help of an example. Since it is pair work, you work with your partner to do the job in turn.

Phrases for reference: find one's application in; be used in; apply sth to; put to use; find utilization in; application of sth. to.

An example for Fig. 3.5 (a): We can find laser application in joining.

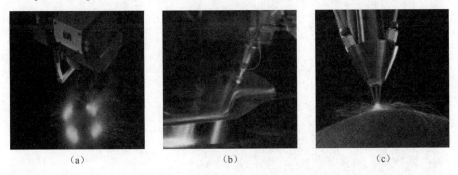

（a）　　　　　　　　　（b）　　　　　　　　　（c）

Fig. 3.5　Figures for expression

（a）Joining; （b）Cutting; （c）Surface treatment

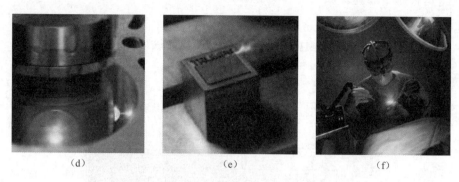

（d）   （e）   （f）

Fig. 3.5   Figures for expression（Continued）
（d）Micro processing；（e）Marking；（f）Surgery

 Text

No one knew what laser would be used for when it was discovered. "A solution waiting for a problem" is a famous saying from the early days of laser. Sometimes, it takes quite a long time before a discovery can be used in a **tangible** way. Yes, it might take 30 years or more before someone realizes that a discovery is really important. Today, laser is used in several areas, such as research, communication, industry, medicine, and **environmental** care.

tangible *adj.* 切实的

environmental *adj.* 周围的，环境的

Industrial Applications of Laser

Today, laser can be found in a broad range of applications within industry, where it can be used for such things as pointing and measuring（Fig. 3.6）. In the manufacturing industry, laser is used to measure the ball **cylindricity** in **bearings** by observing the **dispersion** of a laser beam when reflected on the ball. Yet another example is to measure the shadow of a **steel band** with the help of a laser light to find out the thickness of the band.

cylindricity *n.* 圆柱度
bearing *n.* 轴承
dispersion *n.* 传播，散射
steel band 钢带

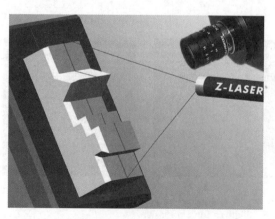

Fig. 3.6   Laser measuring

Within the **pulp** mill industry the concentration of **lye** is measured by observing how the laser beam **refracts** in it.

Laser also works as a **spirit level** and can be used to indicate a flat surface by just **sweeping** the laser beam along the surface. This is, for instance, used when making walls at **building sites**. In the **mining** industry, laser is used to **point out** the drilling direction.

Environmental Applications

Laser technologies have also been used within environmental areas. One example is the ability to determine **from a distance** the environmental **toxins** in a column of smoke. Other examples are being able to predict and measure the existence of **photochemical smog** and **ozone**, both at ground level where it isn't wanted and in the upper **layers** of the atmosphere where it is needed. Laser is also used to **supervise** wastewater **purification**.

Communications

Laser works as a light source in all fiber optics in use (Fig. 3. 7). It has greater **bandwidth** (potentially 100,000 times greater) than an ordinary **copper cable**.

Fig. 3. 7　Laser works as a light source

It is **insensitive** to **interference** from external electrical and magnetic fields. **Crosstalk** (hearing someone else's phone call) is of rare **occurrence**.

Fiber optics is used increasingly often in data and telecommunications around the world.

Research

Laser has become one of the most important tools to researchers within physics, chemistry, biology and medicine throughout the world and is used to:

---

pulp *n.* 果肉，纸浆

lye *n.* 碱液

refract *vt.* 使折射，折射

spirit level 水平仪

sweep *v.* 扫，掠过

building site 建筑工地

mine *vt.* 开矿

point out 指出

from a distance 从远方

toxin *n.* 毒素

photochemical *adj.* 光化学的

smog *n.* 烟雾

ozone *n.* 臭氧

layer *n.* 层，阶层

supervise *v.* 监督；管理；指导

purification *n.* 净化

bandwidth *n.* 带宽

copper cable 铜质电缆

insensitive *adj.* 不敏感的

interference *n.* 干扰

crosstalk *n.* 【电子学】串话；交谈，相声

occurrence *n.* 发生，出现

（1）register **ultra fast chemical processes** such as the **bonding** between **atoms** to form **molecules**;

（2）study the process when cells **split**, or a virus enters into a cell.

The full meaning of this research won't find its answer today but will be shown in the future.

ultra *adj.* 过激的	
ultra fast chemical processes 超快化学过程	
bonding *n.* 连接，结合	
atom *n.* 原子	
molecule *n.* 分子	
split *v.* 劈开，分裂	

 Get to Work

Ⅰ. Matching

Directions：*Match the words or expressions in the left column with the Chinese equivalents in the right column.*

1. optical flats   A. 高真空
2. absorbing crystal   B. 穿透频率
3. digital hologram   C. 最大刻度
4. high vacuum   D. 热扩散
5. false image   E. 光学平面
6. maximum scale   F. 数字全息图
7. arc discharge   G. 误像
8. heat diffusion   H. 声频信号
9. penetration frequency   I. 吸收晶体
10. acoustic signal   J. 电弧放电

Ⅱ. Understanding Checking

Directions：*Fill in the missing parts of the short passage with no more than 3 words based on the above text.*

The invention of laser was a great event in the history, but 1. _____ had passed before its wide use was found.

Laser is being put in use in industry, e. g. it is used to 2. _____ the ball cylindricity and the shadow of a steel band.

Laser is also applied to environmental protection. For instance, it can monitor 3. _____ purification.

Application of laser into communications is one example of its great usage. As it is 4. _____ to negative effects from external electrical and magnetic fields, it is effective in telecommunications.

Research is a field hoping for laser application. Laser has proved its value by means of recording very fast 5. _____ .

III. Passage Completion

Directions: Fill in each of the blanks with one of the words or expressions in the box, making changes if necessary.

beam	achievement	maser	ruby rod
fame	breakthrough	win	electrical engineering

The inventor of the first laser, Theodore Maiman, died on May 5, at the age of 79. Maiman studied 1. _____ at the University of Colorado and Stanford with Willis Lamb, who 2. _____ the Nobel Prize for physics in 1955. Maiman earned his PhD in 1955—two years after Charles Townes invented the first 3. _____, the laser's predecessor generating amplified microwave 4. _____. In May 1960, Maiman made a 5. _____ in the development of the laser. He was able to generate pulses of coherent light by using a 6. _____. In 1984, Maiman received the prestigious Wolf Prize for his 7. _____. In the same year, he became a member of the United States National Inventors Hall of 8. _____.

IV. Phrase Translation

Directions: Complete the sentences by translating into English according to the Chinese given in brackets.

1. _____ (激光用于医疗) to improve precision work like surgery.

2. The coherency (一致性), high monochromaticity (高单色性), and _____ (能产生高能量的特性) are all properties of laser.

3. By _____ (精心设计激光器件), the purity of the laser light can be improved more.

4. In 1917, Albert Einstein in his paper On the Quantum Theory of Radiation (《论辐射的量子性》), _____ (为激光的发明奠定基础).

5. In 1939, Valentin A. Fabrikant predicted the use of stimulated emission to _____ _____ (放大短波).

V. Sentence Remaking

Directions: Simulate the following sentence patterns according to an example provided from the text.

1. It takes a long time before …

E. g. It takes quite a long time before a discovery can be used in a tangible way.

It takes a long time before _____.

2. In …, … is/are used to …

E. g. In the mining industry, laser is used to point out the drilling direction.

In _____, _____ is used to _____.

3. One example is … to …

E. g. One example is the ability to determine from a distance the environmental toxins in a column of smoke.

One example is _____ to _____.

4. ... is/are insensitive to ... from ...

E. g. It is insensitive to interference from external electrical and magnetic fields.

_____ is/are insensitive to _____ from _____.

5. ... won't ... but ...

E. g. The full meaning of this research won't find its answer today but will be shown in the future.

_____ won't _____ but _____.

VI. Para. Translation

Directions: *Translate the following paragraph from English into Chinese.*

A technique that has had recent success is laser cooling. This involves atom trapping, a method where a number of atoms are confined in a specially shaped arrangement of electric and magnetic fields. Shining particular wavelengths of laser light at the ions or atoms slows them down, thus cooling them. As this process is continued, they all are slowed and have the same energy level, forming an unusual arrangement of matter known as a Bose-Einstein condensate.

VII. Mini Imitative Writing

Directions: *The following paragraph originates from the text. Please imitate it and write another one.*

No one knew what laser would be used for when it was discovered. "A solution waiting for a problem" is a famous saying from the early days of laser. Sometimes, it takes quite a long time before a discovery can be used in a tangible way.

3.4   Internet and Lasers

Learning Objectives

In this module, you will:

1. understand ways to connect people to Internet from the sky.

2. learn about lasers' role in delivering the Internet.

3. master strategies adopted by companies in broadening Internet access from the sky.

Warm-up Activity

Directions: After reading the descriptions of Fig. 3.8, please judge they are true or false.

(a)   (b)

Fig. 3.8   Figures for warm-up

(a) The rate of data transmission can be increased in the optical-fiber network;

(b) Laser can be used to make Internet faster

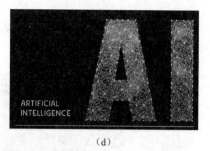

（c）   （d）

Fig. 3. 8   Figures for warm-up （Continued）
（c）Laser marketing can be furthered through online shopping malls；
（d）Laser in combination with AI is the best way for internet development

 Text

Facebook，among other large internet companies，is trying to **figure out** how to connect the next billion people to the Internet.

Without access to expensive **wired** Internet connections，these people largely won't be getting online from computers，but from mobile devices. The race is on to figure out how to connect those devices from the sky—be it by satellite，weather balloon，or **drone**.

In July 2015，Facebook CEO Mark Zuckerberg said he would **envision** a world where people can digitally send our emotions to others. But that technology is still a few years out，so it looks like now Zuck will have to **settle for** drones that shoot Internet through lasers from the sky.

Zuckerberg gave us a **sneak** peak at some of the lasers the Facebook **Connectivity** Team are building in their project to expand Internet connectivity around the world （Fig. 3. 9）.

figure out 解决；算出；想出
wired *adj.* 有线的
drone *n.* 无人机；雄蜂

envision *vt.* 设想

settle for 勉强接受

sneak *adj.* 暗中进行的
connectivity *n.* 连接性，联结性

Fig. 3. 9   Facebook Aquila Plane

Facebook's Connectivity Lab （Fig. 3. 10） is working on a

"laser communication system" that will be able to send data to entire communities. In March 2015, Zuckerberg confirmed that they had completed their first test flight of their drone, which will reportedly have a larger **wingspan** than a Boeing 737 (102 or 138 feet, depending on the model), and weigh less than a car. No photos or videos were released of that drone, but this newest update is the first real look at some of the technology involved in the project.

wingspan *n.* 翼展；翼幅

Fig. 3.10　An employee at Facebook's Connectivity Lab tests the data-delivery laser

The project could change how Internet is **delivered** to those who don't have access to a fixed connection. Combine this laser system with the drone technology the social network is working on, and a picture of a wireless network of internet-connected drones flying overhead starts to **emerge**. The lasers, Zuckerberg noted, will not be **visible** when in use.

deliver *vt.* 交付；发表；递送；释放
emerge *vi.* 浮现；摆脱；暴露
visible *adj.* 明显的；看得见的

Lasers as a form of communication, which **falls under** the **umbrella** of free-space optical communication, is still a relatively new field. NASA reported using a laser to beam data to the moon at 622 Mb/s in 2013, and it's seen some interest from defense **contractors** as well.

fall under 被归入；受到（影响等）
umbrella *n.* 总组织；总称

contractor *n.* 承包人；承包商

Zuckerberg and his team aren't the only ones who want to **broaden** Internet access from the sky—Google's Project Loon looks to use high-**altitude** balloons that emit Internet signals, and even Tony Stark **wannabe** Elon Musk wants to send Internet-beaming satellites to orbit the Earth. And not be **outdone**, Virgin Galactic, led by Sir Richard Branson, announced a large step to sending its own internet satellites to space, **in a partnership with** satellite company OneWeb.

broaden *v.* 变宽，变阔
altitude *n.* 高地；高度
wannabe *adj.* 想要成为的；自封的
outdo *vt.* 超过；胜过

in a partnership with 与……合作

Facebook hasn't given a **timeline** to when the project will be widely available, but Zuckerberg claims that when they are **put into action**, they'll be able to serve 10% of the world's population that is now without Internet **infrastructure**.

Keep dreaming big, Zuck.

timeline *n.* 时间轴, 时间线

put into action 把……付诸实施; 使开始工作

infrastructure *n.* 基础设施; 公共建设

 Get to Work

Ⅰ. Matching

Directions: *Match the words or expressions in the left column with the Chinese equivalents in the right column.*

1. periodic focusing	A. 角色散
2. optical lock-on	B. 闭合共振器
3. argon flash	C. 空对空激光测距
4. excited electron	D. 绝对误差
5. flexible mirror	E. 偏差补偿
6. deviation compensation	F. 光学锁定
7. absolute error	G. 活动反射镜
8. angle dispersion	H. 周期聚焦
9. air-to-air laser ranging	I. 受激电子
10. closed resonator	J. 氩气闪光灯

Ⅱ. Understanding Checking

Directions: *Fill in the blanks with no more than three words according to the text.*

1. It is possible to connect people to the Internet through satellite, weather balloon, or _____ _____.

2. The lasers to expand Internet connectivity around the world are being built by the _____ _____.

3. Facebook's drone will weigh less than _____.

4. The lasers used by Facebook will be _____ when in use.

5. Elon Musk plans to send Internet-beaming _____ to orbit the Earth.

Ⅲ. Passage Completion

Directions: *Fill in each of the blanks with one of the words or expressions in the box, making changes if necessary.*

as	weld	so that	replace
shorten	operation	option	save

Scanner welding is a very fast and dynamic joining technology that shows its strengths in more than just car making. In general, it is an interesting 1. _____ for quantities greater than 100,000 units per year or when many welding 2. _____ are acquired for each component. The time saved by moving faster from one welding spot to the other can 3. _____ the processing time by up to three quarters. A scanner welding station can therefore 4. _____ several conventional welding stations, which not only 5. _____ space but also, as a result of the laser's basic flexibility, allows various components to 6. _____ in one station.

Scanner welding illuminates nonproductive downtime caused by positioning the welding tool. This advantage becomes more significant 7. _____ the distances between the welding spots become greater and the number of weld spots per component increases. There is a positive side effect: without losing any time, the seam sequence can be selected individually 8. _____ the heat input is evenly distributed.

Ⅳ. Phrase Translation

Directions: Complete the sentences by translating into English according to the Chinese given in brackets.

1. This causes rapid heating and _____ (周围空气的爆炸性扩张).

2. The device makes noise similar to _____ (伴随闪电的雷声).

3. When _____ (这种现象发生在某些科学实验), it is referred to as a "plasma mirror" (等离子镜).

4. A laser beam would not _____ (肉眼所不能见) in the near vacuum of space.

5. In 1953, C. H. Townes and others produced the first microwave amplifier, a device that _____ (工作原理与激光类似).

Ⅴ. Sentence Remaking

Directions: Simulate the following sentence patterns according to an example provided from the text.

1. … especially in …, is/are …

E. g. The representation of lasers in popular culture, especially in science fiction and action movies, is often misleading.

_____, especially in _____, is/are _____.

2. It is … to … than …

E. g. It is far easier and cheaper to build infrared laser diodes than visible light laser diodes.

It is _____ to _____ than _____.

3. … to … is likely to …

E. g. Putting enough dust in the air to make the beam visible is likely to be enough to "break" the beam and trigger the alarm.

_____ to _____ is likely to _____.

4. …, while …

E. g. Goldfinger's laser makes a whirring electronic sound, while a real laser produces a fairly heat-free and silent cut.

_____, while _____.

5. … is … as it is/ they are …

E. g. Laser light is not perfectly parallel as it is sometimes claimed.

_____ is _____ as it is/they are _____.

Ⅵ. Para. Translation

Directions: *Translate the follow paragraph from English into Chinese.*

Earlier this month, NASA successfully tested a new technology that allows the agency to beam images, videos, and other data from space to the Earth on lasers rather than the old fashioned way—on radio waves. The benefit of the laser technology is greater transmission speed. The increase could be anywhere from 10 to 1,000 times as fast. "NASA missions collect an enormous amount of data out in space," said Matt Abrahamson, the project manager working for NASA. "Laser communications is a faster alternative for getting those data to the ground."

Ⅶ. Mini Imitative Writing

Directions: *The following paragraph originates from the text. Please imitate it and write another one.*

Facebook hasn't given a timeline to when the project will be widely available, but Zuckerberg claims that when they are put into action, they'll be able to serve 10% of the world's population that is now without Internet infrastructure.

Task Ⅲ　Skill Honing

3.5　Translation of Noun Clauses
名词性从句的翻译

英语的名词性从句包括主语从句、宾语从句、表语从句和同位语从句四种。

一、主语从句的译法

一般采用顺译法和倒译法。

顺译法：按照句子的排列顺序，先译主句，再译主语从句。

It is obvious *that A. C. motors are superior to D. C. motors in many applications*.
显然，交流电机在许多应用场合比直流电机优越。

倒译法：常用于有形式主语 it 的结构中，先译真正的主语从句，再译谓语。

It is a fundamental rule of physics *that energy can neither be created nor destroyed*.
能量既不能被创造也不能被消除，这是物理学的一个基本原则。

二、宾语从句的译法

往往采用顺译法，包括以下两种情况。

1. 动词后的宾语从句。

It's necessary to know *that voltage is used up in components*, *but not in wires*.
有必要知道，电压消耗在元件上，而不是导线上。

2. 介词后的宾语从句。

The scientific research depends on *whether there are proper subjects*.
科学研究取决于是否有合适的实验对象。

三、表语从句的译法

一般可以按原文顺序翻译，多采用"是"的句式。

This is *where the basic principle lies*.
这是基本原理所在。

四、同位语从句的译法

1. 译成定语。

Obviously, there was little probability *that this experiment would succeed*.
很显然，实验成功的可能性极小。

2. 省略连接词前的名词。

The advantage of this device is due to the fact *that it has an automatic temperature control*.
这台设备的优点在于它是自动调温的。

（the fact 省去不译）

3. 译成独立的句子。

We recognize the fact *that the operational amplifier is the most useful single device in analog electronic circuitry*.

我们认同这个事实，运算放大器是模拟电子电路中最有用的独立器件。

Challenge

Directions：*Translate the following sentences into Chinese.*

1. An important consequence of this is that thermal lensing effects inevitably become strong for high output power levels.

2. This is due to the fact that fewer mirrors are required to deliver the laser beam to the cutting head.

3. It is evident that current cannot flow without voltage.

4. Application-based manufacturing differences mean that laser diode and LED products are constructed in fundamentally different ways.

5. It's a good idea that various broadband communication services can be provided.

Task Ⅳ　Job Task Challenges

3.6　Career Speaking: Introducing Enterprises

Ⅰ. Warm Up

Model

(A: manager　　B: guest)

B: Would you like to let me know how many departments you have?

A: Yes, we have 5—Planning Department, Sales Department, Production Department, Personnel Department and *Accounting Department*.

B: I know. You must have a lot of employees.

A: Yes. There are 2,700, including both administrative staff and technical staff. 40% of them have Master's degree.

B: As I know, your employees have good *labor protection* and favorable salary.

A: Right. In addition, we have *recreational facilities*, a medical center and *dormitories* for the single employees.

B: How about their promotion?

A: Every one has an opportunity to be promoted *on the basis of* their performance and *merit*.

B: Your employees are so lucky to work here.

A: All members of the company are part of a large family. Therefore, it is important for us to preserve *harmony* and teamwork.

B: Your words are really impressive!

Notes

Accounting Department	财务部
labor protection	劳动保护
recreational facilities	娱乐设施
dormitory	*n.* 宿舍
on the basis of...	在……的基础上
merit	*n.* 功绩
harmony	*n.* 和睦，和谐

Ⅱ. Matching

Directions: Please choose the corresponding sentences from the box to fill in the blanks. Then, play roles with your partner.

(A: manager　　B: guest)

A: Welcome to our company. Please be seated.

B: Thank you. Your company is very reputed in this city. ＿＿＿＿＿＿＿＿＿＿＿＿＿＿＿

Well, when did it come into being?

A：It was established in 1990. _____

B：Oh, it is still a young company, but your development of optical fiber communications has been remarkable.

A：Indeed. _____ And the demand is getting greater and greater.

B：_____

A：You're right. _____

B：Your company really has strength.

S1：We have been in the trade for more than 20 years.

S2：As far as technology is concerned, I believe our company is among the best in the world.

S3：I can see you are really proud of your company.

S4：I heard much praise to your company.

S5：Our products are well accepted in the world market.

III. Closed Conversation

Directions：Please fill in the blanks according to the Chinese. Then play roles with your partner.

（A：Secretary　　B：Guest）

B：This is my first visit to your company. I'm lucky to have the chance to do so.

A：Welcome. _____ （我现在就带您到四处看看。参观后您会对我们的公司有更深的了解。）

B：I'm sure I'll.

A：This way please. _____ （我们从车间开始吧。）

B：_____ （生产线是全自动的吗?）

A：Yes. _____ （几乎每一道工艺都是由计算机控制的。）

B：It saves quite a lot of manpower.

A：Indeed. _____ （通过自动化我们的效率增加了 30%，成本降低了 15%。）

B：_____ （现在您能给我展示一下如何操作这台机器吗?）

A：No problem. _____ （我需要做的就是按下这儿的电钮。您可以试试。）

B：Very convenient. _____ （产品的销路如何?）

A：_____ （我们的产品近几年销售得很好。而且销量一直在稳步地增长。）

B：_____ （你们有产品目录和价格表吗?）

A：Yes, after the visit I'll give them to you.

B：Thank you. _____（那栋白色的 5 层楼房是干什么的?）

A：_____（在那儿我们有娱乐室、医疗中心和单身宿舍。）

B：Wow! Terrific!

A：Now let's go to the R&D Department.

Ⅳ. Semi-open Conversation

Directions：Please fill in the blanks with your own words. Then play roles with your partner.

（A：Secretary　　B：Guest）

A：Welcome to our company. _____

　　Please ask questions at any time.

B：OK.

A：_____

B：Oh, it is a large company with a long history. _____

A：Yes, we have 7— _____

B：Too many departments! _____

A：We manufacture household appliances.

B：Well, what is your market share?

A：_____

B：It sounds good.

Ⅴ. Live Show

Challenge 1

Directions：Lucy Green, a businesswoman of an American firm, is visiting a Chinese company that makes and sells optical instruments in case there is possibility to establish a long-term cooperation. Wang Fang, is in charge of showing her around. Please act it out with your partner or partners. The model and the useful expressions below are just for your reference.

Challenge 2

Directions：The students in groups discuss how to give a general picture of a company including its history, size, employment, management, business scope, product quality, annual output, production process and other aspects. One student of each group makes an oral report to the whole class. The model and the useful expressions below are just for your reference.

Model

（A：Secretary　　B：Mr. Smith）

A：Welcome, Mr. Smith. I'll show you around. I am sure you will know our company better after the visit.

B：That'll be more helpful.

A：If you have any questions, please feel free to ask. I will answer any questions that I can, except for those concerning corporate secrets.

B：OK.

A：Our site covers an area of approximately 30,000 m^2 and we have about 1,000 employees.

B：It's much larger than I expected.

A：Do you see that white building on your right?

B：Yes.

A：That's our *office block*. We have four departments there：Advertising, Purchasing, Marketing and *R&D*.

B：What sort of products do you deal with?

A：We *specialize in* manufacturing *a wide range of* telecommunication *devices* and export them all over the world.

B：Good. Where is your factory?

A：Mr. Smith, this way please. On the opposite of the office block, that gray house of three storeys is our factory. All parts are manufactured and *assembled* there.

B：It looks rather *magnificent*! Does the factory work with everything from the *raw* material to the finished product?

A：Yes. We've introduced new technology and automation system, so the whole production takes about only 25 minutes from beginning to end.

B：An efficient factory! How do you ensure quality control?

A：All products have to pass strict inspection before they go out. We control quality by computer. And of course, the final inspection is done by the engineers.

B：Would you mind telling me the *annual output value* approximately?

A：No. The annual output value varies between 7 – 8 million RMB.

B：Terrific! Do you spend a lot on research and development each year?

A：Yes, we invest almost one third of the profits in it.

B：Would it be possible for me to have a closer look at your samples?

A：No problem. I'll take you down to our showroom. Follow me, please.

B：I'd like to have your latest *catalogues* or something that tells me about your company and your products.

A：OK. Here you are.

Notes

office block	办公大楼
R&D	*abrr.* 研究开发（research and development）
specialize in	专攻，专门研究
a wide range of	各式各样的
device	*n.* 装置，设备
assemble	*v.* 装配
magnificent	*adj.* 壮观的
raw	*adj.* 未加工的

| annual output value | 年产值 |
| catalogue | *n.* 目录 |

Useful Expressions

（1）客户咨询。

How many departments do you have in your company?

What are your company's main products?

What was the total amount of your sales last year?

（2）接待来访。

We are honored to show our company to such a distinguished group of guests.

I will give you a complete picture of our operation.

Maybe we could start with the Designing Department.

（3）公司概况。

We now command 35% of the market in home appliances.

The company was established in 1965, and we have about 5,500 employees now.

Our company is one of the world's biggest manufacturers of optical communication receivers, with over 30 branches at home and abroad.

（4）产品说明。

There is a good market for these articles.

One of the main strengths is the quality of our products.

All products have to pass strict inspection before they go out.

（5）生产过程。

We've increased our efficiency by 20% through automation.

Production takes about 25 minutes from beginning to end.

What the workers should do is to push the button here.

3.7　Writing：Letter of Congratulation

　　在西方，当得知有亲友或同事要订婚、升迁、获奖等喜庆之事时，往往会写信表示祝贺。内容通常包含三个方面：①直接点明事件，表达喜悦之情；②对事件进行积极评价，赞扬、祝福对方；③表示真诚的祝贺与期望。此外，凡是遇到重大的喜庆节日，如圣诞节、新年等，人们也会写祝贺信，但语言会更简洁，寥寥数语即可。祝贺信传达的是美好温馨的情感，因此语言应热情洋溢，亲切自然。

Sample 1

Dear Sue,

　　We are very glad to hear the news of your promotion to Sales Manager of the company. Congratulations!

　　It is a well-deserved ***recognition*** of your hard work and great achievements. We're encouraged by your efforts rewarded and we are proud of you.

We send you our best wishes for success and happiness in your future.

Yours cordially,
Helen

Note

recognition   *n.* 认可

Sample 2

Dear Mary,

I offer you my warmest congratulations on your ***engagement***. That's great news about you and your ***fiancé*** Bob! I know you have common interests and tastes in every way. I send you both every good wish.

May the years ahead bring you joy and ***contentment***!

Yours,
Wang Xiao

Notes

engagement	*n.* 订婚
fiancé	*n.* ［法］未婚夫
contentment	*n.* 满足

Useful Expressions

1. I am so pleased to hear that…

2. I offer you my warmest congratulations on your…

3. Heartfelt congratulations on…

4. Accept our heartiest congratulations and all good wishes.

5. Please accept our sincerest congratulations and very best wishes for all the good future.

Challenge

Directions：*Please write a letter of congratulation to your colleague Susan, who has been rewarded due to her outstanding achievements.*

Task V   Output and Evaluation

Laser Construction					
Target	Understand and describe laser construction				
Requirement	Make an English presentation according to lesson 3. 2. Draw a mind map. Produce micro-video to express professional knowledge. Add accurate Chinese and English subtitles to the video.				
Contents	Laser Construction： Pump Source Gain Medium Optical Resonator				
Group： _____	**Project**	**Name**	**Software**	**Score**	**Requirements**
	Mind map				The logic should be sound, and the keywords used should be accurate.
	Script				The format should adhere to the specified guidelines, and it should cover all necessary contents.
	PPT				It should be consistent with the logic presented in the mind map. Clear and concise content/Consistent formatting/Limited text/Engaging graphics and animations.
	Speaking				Clear/fluent
	Subtitle				They should be bilingual, correct and synchronized with the student's speech.
	MP4				The video should effectively convey knowledge, be accurate, and visually engaging.
Operating environment	Win7/Win8/Win9/Win10/Win11/Mobile phone				
Product features					

Laser Construction
思维导图

Laser Construction
微课

Script reference

Laser Construction

标题	专业内容（讲稿与中英字幕）	PPT 中需显示的文字	表现形式	素材
每个知识点的要点，或者小节名字	做报告时候的英文讲稿 视频中的中、英字幕	需要特别提醒的文字	讲解到某个地方或某句话时，出现的图片、动画或者文字	图片或者视频
	Laser is constructed from three principal parts: An energy source (usually referred to as the pump or pump source), a gain medium or laser medium, and two or more mirrors that form an optical resonator. 激光是由三个主要部分组成：能源（通常被称为泵或泵源）、增益介质或激光介质，两个或两个以上的镜子组成的光学谐振腔。		主要内容和激光器结构图	图 1 激光器的结构 Nd:YAG solid-state laser Highly reflective mirror Flashlamp(pump source) Partially reflective mirror ND:YAG crystal(laser medium) Laser output Optical resonator
Contents	**Laser Construction** Contents： 1. Pump source 2. Gain medium 3. Optical resonator **激光器的结构** 内容： 1. 泵浦源 2. 增益介质 3. 光学谐振腔	**Laser Construction** Contents： 1. Pump source 2. Gain medium 3. Optical resonator		
1. Pump source 泵源	The pump source is the part that provides energy to the laser system. 泵源是向激光系统提供能量的部分。		强调图中泵源部分	图 1 激光器的结构
	Examples of pump sources include electrical discharges, flashlamps, arc lamps, light from another laser, chemical reactions and even explosive devices. 泵源是向激光系统提供能量的部分，如放电、闪光灯、弧光灯，另一激光器的光束，化学反应，甚至爆炸装置。	Examples of pump sources： Electrical discharges Flashlamps Arc lamps Light form another laser Chemical reactions Explosive devices	依次出对应的图	图 2 电子放电

续表

标题	专业内容（讲稿与中英字幕）	PPT 中需显示的文字	表现形式	素材
1. Pump source 泵源				  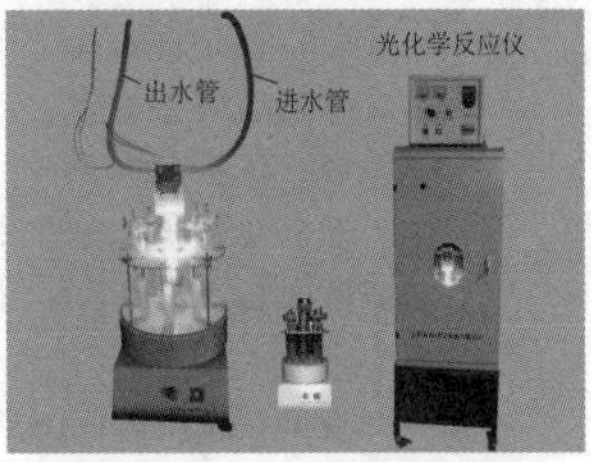 图3　闪光灯  

图4　弧光灯

图5　另一激光器的光束

图6　光化学反应

图7　爆炸

标题	专业内容（讲稿与中英字幕）	PPT 中需显示的文字	表现形式	素材
1. Pump source 泵源	The type of pump source used principally depends on the gain medium, and this also determines how the energy is transmitted to the medium. 泵源的类型主要取决于增益介质。这也决定了能量如何传送给介质。 A helium-neon laser uses an electrical discharge in the helium-neon gas mixture, a Nd：YAG laser uses either light focused from a xenon flash lamp or diode lasers, and excimer lasers use a chemical reaction. 氦氖激光器采用氦氖混合气体放电；Nd：YAG 激光器使用来自氙气闪光灯或二极管激光器的光束；受激准分子激光器则使用化学反应。	A helium-neon laser uses an electrical discharge in the helium-neon gas mixture. A Nd：YAG laser uses either light focused from a xenon flash lamp or diode lasers. Excimer lasers use a chemical reaction.	文字介绍 3 种激光器的泵源	
2. Gain medium 增益介质	The gain medium is the major determining factor of the wavelength of operation, and other properties of the laser 增益介质是激光工作波长及其他特性的决定性因素。	The gain medium is the major determining factor of the wavelength of operation, and other properties of the laser.	强调图中增益介质部分	图1
2. Gain medium 增益介质	The gain medium is excited by the pump source to produce a population inversion, and it is in the gain medium that spontaneous and stimulated emission of photons takes place, leading to the phenomenon of optical gain, or amplification. 泵源激活增益介质，产生粒子数反转，在增益介质中，会发生光子的自发和受激发射，导致光学增益现象，即放大。	The gain medium is excited by pump source to produce：population inversion. Spontaneous and stimulated emission of photons in gain medium, or leads to：optical gain amplification.	强调图中增益介质部分	图1
Examples of different gain media 增益介质	Examples of different gain media include： 不同增益介质包括：			
Examples of different gain media 增益介质	Liquids, such as dye lasers. The exact chemical configuration of the dye molecules determines the operation wavelength of the dye laser. 液体，如染料激光器。染料分子的精确化学构造决定染料激光的工作波长。	Liquids, such as dye lasers.	液体图	图8　液体 

续表

标题	专业内容（讲稿与中英字幕）	PPT 中需显示的文字	表现形式	素材
Examples of different gain media 增益介质	Gases, such as carbon dioxide and mixtures such as helium-neon. These lasers are often pumped by electrical discharge. 气体，如二氧化碳和氦氖混合物。这类激光器通常通过放电抽运。	Gases, such as carbon dioxide and mixtures such as helium-neon	气体图	图9　气体
	Solids, such as crystals and glasses. The solid host materials are usually doped with impurities. 如晶体和玻璃等固体。固体基质材料通常掺杂杂质。	Solids, such as crystals and glasses	晶体和玻璃	图10　晶体
	Semiconductors, a type of solid, in which the movement of electrons between material with differing dopant levels can cause laser action. 半导体是固体的一种，电子在不同掺杂水平的材料间的流动，能产生激光作用。	Semiconductors, a type of solid	半导体	图11　半导体
3. Optical resonator 光学谐振器	The optical resonator, or optical cavity, in its simplest form is two parallel mirrors placed around the gain medium which provide feedback of the light. 光学谐振器或光学腔的最简单的形式是在增益介质周围平行放置两个反射镜，反馈光束。	The optical resonator, or optical cavity, in its simplest form is two parallel mirrors placed around the gain medium which provide feedback of the light.	强调图中光学谐振腔	图1　激光器的结构
	The mirrors are given optical coatings which determine their reflective properties. Typically, one will be a high reflector, and the other will be a partial reflector. The latter is called the output coupler, because it allows some of the light to leave the cavity to produce the laser's output beam. 反射镜的镀膜决定其反光性能。通常一面是全反射镜，另一面是部分反射镜。后者被称为输出耦合器，因为它允许部分光，离开腔，输出激光光束。	Typically, one will be a high reflector, and the other will be a partial reflector.	强调两块镜子	图1　激光器的结构

标题	专业内容（讲稿与中英字幕）	PPT 中需显示的文字	表现形式	素材
3. Optical resonator 光学谐振器	In more complex lasers, configurations with four or more mirrors forming the cavity are used. The design and alignment of the mirrors with respect to the medium is crucial to determining the exact operating wavelength and other attributes of the laser system. 在更为精密的激光器，谐振腔由4个或更多的反射镜构成。与介质相关的反射镜的设计和排列决定激光系统的精确工作波长及其他特性。	In more complex lasers, configurations with four or more mirrors forming the cavity are used.	谐振腔由4个或更多的反射镜构成	 图12　多个反射镜 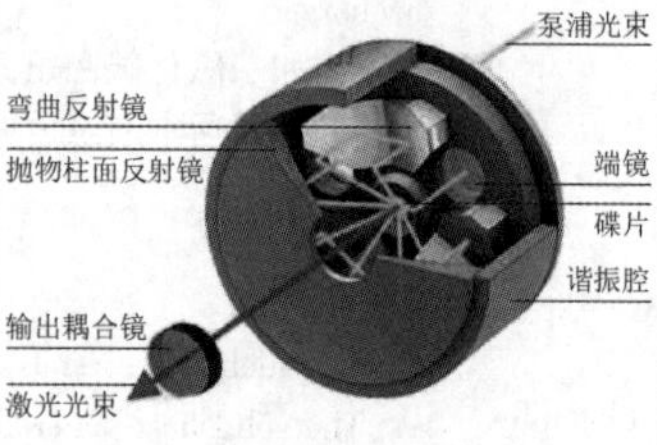
	Light from the medium, produced by spontaneous emission, is reflected by the mirrors back into the medium, where it may be amplified by stimulated emission. The light may reflect from the mirrors and thus pass through the gain medium many hundreds of times before exiting the cavity. 由自发辐射产生的介质光经镜子反射重回到介质，经受激发射后可以被放大。光能被镜子反射，因此在溢出腔外前数百次地穿越增益介质。	Light from the medium, produced by spontaneous emission, is reflected by the mirrors back into the medium, where it may be amplified by stimulated emission. The light may reflect from the mirrors and thus pass through the gain medium many hundreds of times before exiting the cavity.	强调光在溢出腔外前数百次地穿越增益介质	 图13　光路 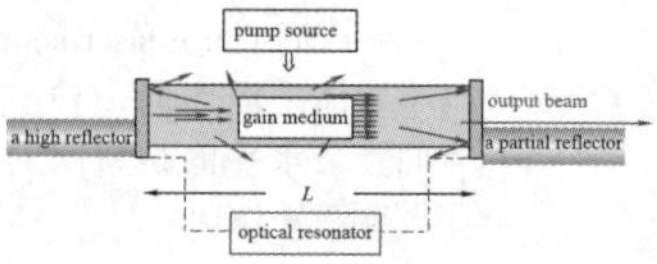
	The end. Thank you!	The end. Thank you!		

Task Ⅵ　Knowledge Expansion

3.8　A Prospectus about Laser Machines

TruLaser Cell Series 3000

It is the flexible, complete solutions for small components.

The TruLaser Cell Series 3000 (Fig. 3. 11) offers a flexible solution for 2D and 3D processing of small parts. Typical applications are to be found in the field of medical technology, in precision mechanics and in the electronic industry—wherever the emphasis is on high precision, quality and safety.

Fig. 3. 11　TruLaser Cell Series 3000

LASERCELL 1005

It is for the entire world of flexible laser material processing.

Whether you have a job shop with constantly changing tasks, or need to meet strict requirements as in the aerospace industry or serial production in the automotive industry, the LASERCELL 1005 (Fig. 3. 12) unites flexibility and productivity—also in the rough industrial environment.

Fig. 3. 12　LASERCELL 1005

LASERCELL 1000

It is indispensable for round parts.

Radial seams, axial seams and everything else in between—the LASERCELL 1000 (Fig. 3. 13) is geared for these applications. On this compact machine, the laser, the electrical parts and fixtures share a common base plate. An extensive range of accessory equipment is available, as well as clearly defined interfaces to all conventional quality assurance systems.

Fig. 3. 13　LASERCELL 1000

LASERCELL 6005

It is flexible, also for extra large parts.

The LASERCELL 6005 (Fig. 3. 14) offers just the solution for the laser processing of work pieces as big as automobiles. Whether you are laser cutting or laser welding, the system enables cost-efficient operation coupled with full flexibility in terms of laser power and automation components.

Fig. 3. 14　LASERCELL 6005

TruLaser Robot 5020

It is a module for complex work pieces.

The TruLaser Robot 5020 (Fig. 3. 15) is a complete robot cell for 3D welding of complex work pieces and mass production. A flexible laser light cable guides the laser beam to the processing optics at the robot arm.

Fig. 3. 15   TruLaser Robot 5020

Exercise

Which can do what?

Directions：Please link the laser on the left with its corresponding functions and features on the right by drawing a line according to the text.

Lasers
TruLaser Cell Series 3000
LASERCELL 1005
LASERCELL 1000
LASERCELL 6005
TruLaser Robot 5020

Functions and Features
complex work pieces
axial seams made
coping with extra large parts
3D processing of small parts
a shared base plate
cost-efficient operation
help of the robot arm
working in rough industrial environment
applications in precision mechanics
serial production in the auto industry

3. 9   Using Lasers

The use of lasers in medicine is complex, and your facial plastic surgeon is trained in the use of lasers and understands how and when to use a laser. Your surgeon will decide if a laser is appropriate, and which laser is best for the situation.

In medicine, physicians can use lasers to make incisions, vaporize tumors, close blood vessels, or even treat skin wrinkles. The laser makes it possible to change tissue without making an incision. So a surgeon can treat birthmarks or damaged blood vessels, remove port-wine stains (葡萄酒色痣), and shrink facial "spider veins" (蜘蛛状血管病) without major surgery.

Is it any wonder that many facial plastic surgeons use lasers on a routine basis? They use the laser as a "light scalpel" (解剖刀). The tissue is left sterile, and bleeding is greatly reduced. When the laser is used to treat port-wine stains, no cuts are made. The laser energy penetrates through the skin to shrink the abnormal blood vessels that are the cause of these marks.

There are different types of laser surgery, e. g. laser skin peeling. Lasers can be used to reduce wrinkles around the lips or eyes, even the entire face, softening fine wrinkles and removing certain blemishes (瑕疵) on the face.

Another example is laser removal of birthmarks and skin lesions （损害）. Port-wine stain birthmarks respond remarkably well to laser treatment. The abnormal blood vessels that cause these marks are reduced in size by the laser. This results in a lightening of the treated area. Skin growths, facial "spider veins", and some tattoos respond to laser surgery. Most situations take more than one laser treatment, but some respond to a single treatment.

The facial plastic surgeon often uses the minimum laser intensity. The low intensity requires many treatments. However, the low intensity also preserves as much of the healthy tissue as possible. This produces an aesthetically pleasing result. Many of these laser surgeries are performed as outpatient treatments in hospitals or offices.

After your surgeon has indicated that a laser can be helpful in the surgery, your surgeon will explain the laser of choice and what can be accomplished. As with all surgery, the laser has its limitations. Often the results are spectacular. Your surgeon will give you the best judgment for the particular procedure.

Some surgeons may use local anesthetics （麻醉剂） to numb the treated area before the surgery. Surgery can sometimes be done in the surgeon's office; other times the surgeries are performed in outpatient facilities at a clinic or hospital. Your surgeon will decide on the appropriate method, dictated by the nature of the surgery.

Because safety is a major component of laser use, your surgeon will describe safety precautions before the surgery. If you are treated with a local anesthetic, you will be required to wear protective glasses or goggles during laser use.

Exercises

Directions: Choose the best answers from the four choices.

1. What is the meaning of the word "wrinkle" in the second line of Para. 2?

A. line   B. scar

C. color   D. shortcoming

2. How are port-wine stains treated through a laser?

A. A laser makes an incision.

B. Laser skin peeling is a good solution.

C. Some cuts are made by a laser.

D. The laser passes into the skin to treat the abnormal blood vessels.

3. Which of the following statements is false according to the passage?

A. Laser skin peeling can remove certain flaws on the face.

B. The laser works in reducing the marks that cause the abnormal blood vessels.

C. The laser plays a part in treating some tattoos.

D. A single laser treatment is enough in most cases.

4. Why is low laser intensity often used in facial plastic surgery?

A. Low laser intensity is usually demanded by patients.

B. The surgeon wants to make the operation cost-effective.

C. Low laser intensity can lead to be healthier and better looking.

D. Its effectiveness is a determining factor.

5. Which one is true about local anesthetics during laser use?

A. Local anesthetics are necessary in most operations.

B. With a local anesthetic, the treated area is numbed.

C. Some risks about local anesthetics are not included in the surgeon's precaution before the surgery.

D. Protective glasses or goggles have nothing to do with a local anesthetic.

Unit 4   Optical Detection

In this unit, you are going to learn the following contents:

Task Ⅰ   Technical Principles and Device Cognition

4.1   Photoelectric Effect

4.2   Photodetector

Task Ⅱ   Technology Application Understanding

4.3   Optical Detection Techniques

4.4   CCD

Task Ⅲ   Skill Honing

4.5   Translation of Attributive Clauses

Task Ⅳ   Job Task Challenges

4.6   Career Speaking: Product Introduction

4.7   Writing: E-mail

Task Ⅴ   Output and Evaluation

Task Ⅵ   Enhancing Competency

4.8   LED Technology

4.9   Optical Storage Technology

Unit 4

Optical Detection

Task I   Technical Principles and Device Cognition

4.1   Photoelectric Effect

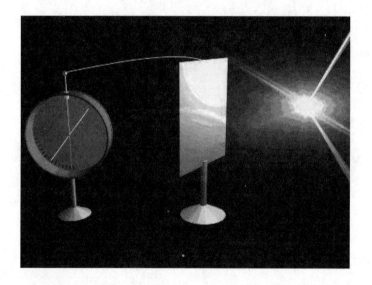

 **Learning Objectives**

In this section, you will:

1. understand the definition of photoelectric effect.

2. comprehend the experiment: no matter how bright a red light you have, it will not produce a current in a metal, but even a very dim blue light will result in a current flowing.

A: these results can not be explained if light is thought of as a wave, why? (amount of energy: brightness and wave)

B: with light as photon, these results can be explained. Try to explain.   (energy of individual photon)

3. understand that a good way to think of the photoelectric effect is like a full car park with lots

of really bad drivers. Relationship between car park model and photoelectric effect. （electron /red photon/blue photon）

 Warm-up Activity

Do you know names of the applications based on photoelectric effect in Fig. 4. 1？

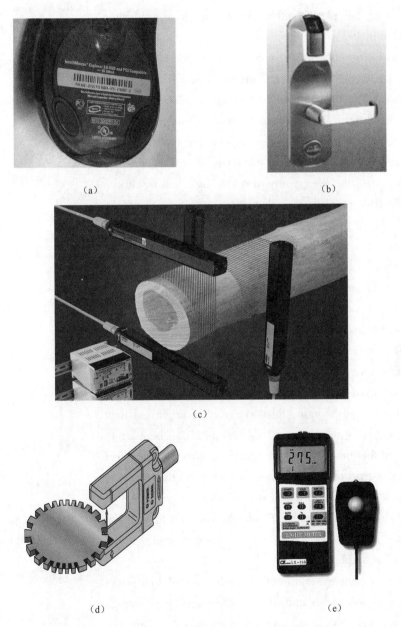

(a)　　　　　　　　　　　　(b)

(c)

(d)　　　　　　　　　　　　(e)

Fig. 4. 1　Applications based on photoelectric effect

Text

In 20th century's physics, two ideas stand out as being totally revolutionary: **relativity** and **quantum theory**. Although Einstein is best known for his theory of relativity, he also played a major role in developing quantum theory. And it was his contribution to quantum theory—explaining the **photoelectric effect**—which won Einstein his **Nobel Prize** in 1921.

relativity *n.* 相对性，相关性，［物］相对论
quantum theory 量子论
photoelectric effect 光电效应
Nobel Prize 诺贝尔奖

The photoelectric effect is the name given to the **observation** that when light is shone onto a piece of **metal**, a small current flows through the metal. The light is giving its energy to electrons in the atoms of the metal and allowing them to move around, producing the current. However, not all colors of light affect metals in this way. No matter how bright a red light you have, it will not produce a current in a metal, but even a very **dim** blue light will result in a current flowing. The problem was that these results can not be explained if light is thought of as a wave. Waves can have any amount of energy you want—big waves have a lot of energy, small waves have very little. And if light is a wave, then the **brightness** of the light affects the amount of energy—the brighter the light, the bigger the wave, the more energy it has. The different colors of light are defined by the amount of energy they have. If all else is equal, blue light has more energy than red light with yellow light somewhere in between. But this means that if light is a wave, a dim blue light would have the same amount of energy as a very bright red light. And if this is the case, then why won't a bright red light produce a current in a piece of metal as well as a dim blue light? Einstein realized that the only way to explain the photoelectric effect was to say that **instead of** being a wave, as was generally accepted, light **was** actually **made up of** lots of small packets of energy called photons that behaved like particles. Einstein wasn't the first person to use the idea of photons, but he was the first to make it the starting point of an **explanation**.

observation *n.* 观察，观测，观察资料（或报告）
metal *n.* 金属
dim *adj.* 暗淡的，模糊的，无光泽的
brightness *n.* 光亮，明亮，聪明
instead of 代替，而不是……
be made up of 由……组成
explanation *n.* 解释，解说，说明，辩解，互相讲明
dislodge *v.* 驱逐

With light as photons, Einstein showed that red light can't **dislodge** electrons because its individual photons don't have enough energy—the impacts are just not large enough to shift the electrons. However, blue light can dislodge electrons—each **individual** photon has more energy than the red photon. And

individual *adj.* 个别的，单独的，个人的

112

photons of **ultraviolet light**, which have yet more energy, will give electrons enough energy to **whizz** away from the metal altogether. A good way to think of the photoelectric effect is like a full car park with lots of really bad drivers. There is a car parked in a space, and lots of other drivers want that space. To get it, they can try knocking the parked car out of the way, but they can only manage to hit it one car at a time. A tiny red mini just won't have the energy to knock the parked car out of the parking space, but a big blue van will. And imagine hitting the parked car with a big ultraviolet **lorry**—the parked car is most likely going to move far enough to **collide** with something else. Returning to light and electrons, there is never really just one photon of light at a time. A bright light **emits** lots of photons, but it doesn't matter how bright a red light gets; red photons will still not be able to **budge** a single electron. This is like having a car park full of red minis each **randomly** hitting a parked car in turn—there will be a lot of **dents** but the parked car will remain where it is. However, even a dim blue light will **shift** some electrons—we know that even one blue van will be able to move the parked car.

Einstein's explanation of the photoelectric effect was just the start of an **avalanche** of discoveries that became quantum theory. In this theory, light is not just a particle and not just a wave: it can be one or the other, depending on how it is measured. And it was discovered later that even the electrons are not just particles but are waves too.

ultraviolet light 紫外（射）线

whizz *v.* （使）飕飕作声

lorry *n.* ＜古＞卡车，铁路货车

collide *vi.* 碰撞，抵触

emit *vt.* 发出，放射，吐露，散发，发表，发行

budge *v.* 移动

random *adj.* 任意的，随便的，胡乱的

dent *n.* 凹，凹痕，（齿轮的）齿，弱点

shift *v.* 替换，转移，改变，移转，推卸，变速

avalanche *n.* 雪崩

particle *n.* 粒子，点，极小量，微粒

 Get to Work

I . Matching

Directions: Match the words or expressions in the left column with the Chinese equivalents in the right column.

Photoelectric Effect
（car park model）
动画

1. relativity
2. quantum theory
3. ultraviolet light
4. electromagnetic wave
5. frequency and period

A. 频率与周期
B. 紫外光
C. 电磁波
D. 量子理论
E. 相对论

II. Understanding Checking

Directions: Mark Y (for YES) if the statement agrees with the information given in the text; N (for NO) if the statement contradicts the information given in the text.

(   ) 1. The quantum theory is the name given to the observation that when light is shone onto a piece of metal, a small current flows through the metal.

(   ) 2. Although Einstein is best known for his theory of relativity, he also played a major role in developing quantum theory.

(   ) 3. No matter how bright a blue light you have, it will not produce a current in a metal, but even a very dim red light will result in a current flowing.

(   ) 4. However, blue light can dislodge electrons — each individual photon has more energy than the red photon.

(   ) 5. It was discovered later that even the electrons are not just particles but are waves too.

III. Passage Completion

Directions: Fill in each of the blanks with one of the words or expressions in the box, making changes if necessary.

finally	begin	more	red	turn
observe	once	measure	through	respect

The experiment itself is now performed in appropriate physics classes at most colleges and universities, and possibly in some high schools. It isn't difficult to perform, and if proper care is used, will give quite accurate results.

We 1. _____ with a photoelectric tube (or phototube), which is a vacuum tube containing a metal plate curved into a half-cylinder (the anode) and a thin wire electrode (the cathode) along the axis of the cylinder. The figure to the right shows this as seen from the top. We use a very sensitive meter, called a galvanometer (G), to 2. _____ the current passing through the tube, and a variable voltage source (V) to control the voltage applied between the two electrodes inside the tube. 3. _____, we add a light source (usually a mercury vapor lamp, but it can be an arc light or even a tungsten filament light bulb), and a colored filter to limit the light striking the photoelectric tube to a single frequency.

With no light, of course, the current through the tube will be zero, regardless of the applied voltage. (Yes, there is a slight capacitance between the metal plate and the wire, and a high enough voltage will cause an arc. But the capacitance is very slight and will charge quickly, and we won't be using anywhere near enough voltage to cause any problems.) So we 4. _____ on the light, place one of our colored filters between the light and the phototube, and begin our measurements.

The first thing we note is that if we reverse the polarity of the voltage source V, increasing the voltage or the intensity of the light will increase the current flow. This seems logical; the light is "kicking" electrons off the metal plate, and they are attracted to the now positively-charged wire electrode. The 5. _____ light, the more electrons and hence the higher the current.

However, when we make the wire electrode negative with 6. _____ to the metal plate as shown in this figure, we begin to 7. _____ some interesting effects. First, 8. _____ light, infrared (IR) radiation and anything of a lower frequency will not excite the phototube enough to enable current to flow. You pretty much have to get up at least to green light before you can measure a reaction. (This may vary depending on the specific metal coating the surface of the curved electrode.)

9. _____ you find a filter whose color allows current to flow 10. _____ the tube, you can begin to take your measurements.

IV. Translation

Directions: Complete the sentences by translating into English according to the Chinese given in brackets.

1. In 1887, the German physicist Heinrich Rudolf Hertz discovered an _____ _____ (物质的有趣的性质).

2. The main relationship developed in this chapter was the one between the incident optic power and _____ (光电检测器中产生的电流).

3. With no light, of course, the current through the tube will be zero, _____ _____ (不考虑外加电压).

4. The different colors of light _____ (根据……定义的) by the amount of energy they have.

5. The first thing we note is that if we _____ (改变电压源的极性), increasing the voltage or the intensity of the light will increase the current flow.

4.2　Photodetector

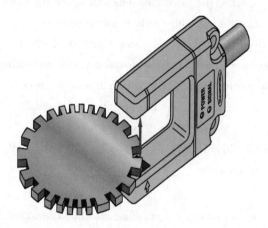

Learning Objectives

In this module, you will:

1. understand the meaning of the word "photosensor".

2. learn about photodetector types.

3. learn about photodetector properties.

Warm-up Activity

Do you know names of the applications based on photoelectric effect in Fig. 4.2?

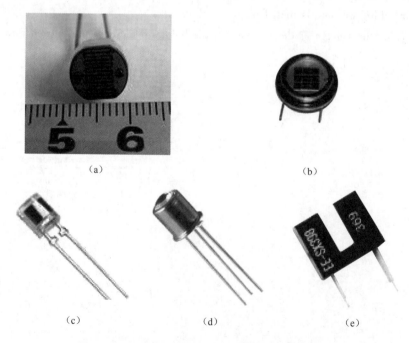

Fig. 4.2　Applications based on photoelectric effect

Text

Photosensors or **photodetectors** are **sensors** of light or other electromagnetic energy. A photodetector has a PN junction that converts light photons into current. The absorbed photons make **electron-hole pairs** in the **depletion region**. Photodiodes and photo transistors are a few examples of photo detectors. Solar cells convert some of the light energy absorbed into electrical energy.

Types

Photodetectors may be classified by their mechanism for detection.

Photoemission or photoelectric effect: Photons cause electrons to transition from the conduction band of a material to free electrons in a vacuum or gas.

Thermal: Photons cause electrons to transition to mid-gap states then decay back to lower bands, inducing phonon generation and thus heat.

Polarization: Photons **induce** changes in polarization states of suitable materials, which may lead to change in index of refraction or other polarization effects.

Photochemical: Photons induce a chemical change in a material.

Weak interaction effects: Photons induce secondary effects such as in photon drag detectors or gas pressure changes in Golay cells.

Photodetectors may be used in different configurations. Single sensors may detect overall light levels. A 1-D array of photodetectors, as in a spectrophotometer or a line scanner, may be used to measure the distribution of light along a line. A 2-D array of photodetectors may be used as an image sensor to form images from the pattern of light before it.

A photodetector or array is typically covered by an illumination window, sometimes having an anti-reflective coating.

Properties

There are a number of **performance metrics**, also called figures of merit, by which photodetectors are characterized and compared.

Spectral response: The response of a photodetector as a

photosensor *n.* 光敏元件，光传感器

photodetector *n.* 光电探测器；光检测器

sensor *n.* 传感器，灵敏元件

electron-hole pair 电子空穴对

depletion region 耗尽区

thermal *adj.* 热的，保热的；温热的；*n.* 上升的暖气流

polarization *n.* 极化；产生极性；（光）偏振；对立

induce *vt.* 引起；归纳；引诱；［电］感应

photochemical *adj.* 光化学的

weak interaction effect 弱相互作用效应

performance metric ［计］性能度量

special response 光谱响应

function of photon frequency.

Quantum efficiency：The number of carriers（electrons or holes）generated per photon.

Responsivity：The output current divided by total light power falling upon the photodetector.

Noise-equivalent power：The amount of light power needed to generate a signal comparable in size to the noise of the device.

Detectivity：The square root of the detector area divided by the noise equivalent power.

Gain：The output current of a photodetector divided by the current directly produced by the photons incident on the detectors, i. e. , the built-in current gain.

Dark current：The current flowing through a photodetector even in the absence of light.

Response time：The time needed for a photodetector to go from 10% – 90% of final output.

Noise spectrum：The intrinsic noise voltage or current as a function of frequency. This can be represented in the form of a noise spectral density.

Nonlinearity：The RF-output is limited by the nonlinearity of the photodetector.

quantum efficiency 量子效率

responsivity *n.* 响应率

noise-equivalent power ［医］噪声等效功率

detectivity *n.* 探测灵敏度

gain *n.* 增益

dark current 暗流；暗电流，无照电流

response time 响应时间

noise spectrum 噪声谱

nonlinearity *n.* 非（直）线性（特性）

 Get to Work

Ⅰ. **Matching**

Directions：*Match the words or expressions in the left column with the Chinese equivalents in the right column.*

1. potential barrier   A. 辐射通量

2. electron-hole pair   B. 光电池

3. cavity electron hole   C. 光敏电阻

4. absorption ratio   D. 电子空穴对

5. visible light   E. 势垒

6. photoelectric triode   F. 空穴

7. multiplier phototube   G. 可见光

8. photoelectric cell   H. 光电倍增管

9. photosensitive resistor   I. 吸收率

10. radiant flux   J. 光电三极管

II. Understanding Checking

Directions: Mark Y (for YES) if the statement agrees with the information given in the text; N (for NO) if the statement contradicts the information given in the text.

(   ) 1. Having no free charges, its resistance is low; as a result, almost all the voltage drop across the diode appear across the junction itself.

(   ) 2. When reverse biased, the potential energy barrier between the P and N regions increases.

(   ) 3. To increase the response, a preamplifier may be integrated onto the same chip as the diode. The resulting device is an integrated detector preamplifier (IDP).

(   ) 4. For emission, the diode is reverse biased, and charges injected into the junction recombine to produce photons.

(   ) 5. For detection, the diode is reverse biased, and incoming photons generate electron-hole pairs, producing electrical current.

III. Passage Completion

Directions: Fill in each of the blanks with one of the words or expressions in the box, making changes if necessary.

processing	slow	systems	before	wavelengths

Light can be detected by the eye. The eye is not suitable for modem fiber communications because its response is too 1. _____, its sensitivity to low-level signals is inadequate, and it is not easily connected to electronic receivers for amplification, decoding, or other signal 2. _____. Furthermore, the spectral response of the eye is limited to 3. _____ between 0. 4 and 0. 7 μm, where fibers have high loss. Nonetheless, the eye is very useful when fibers are tested with visible light. Breaks and discontinuities can be observed by viewing the scattered light. 4. _____, such as couplers and connectors, can be visually aligned with the visible source 5. _____ the infrared emitter is attached.

IV. Translation

Directions: Translate the following paragraph from English into Chinese.

The vacuum photodiode and photomultiplier tube are not placed in operational fiber communications systems, although they can be useful in testing of fiber components. The high sensitivity of the photomultiplier makes it particularly helpful when measuring low levels of optic power. Photoemissive detectors are somewhat easier to explain than semiconductor devices, and the two have many properties in common.

Task Ⅱ   Technology Application Understanding

4.3   Optical Detection Techniques

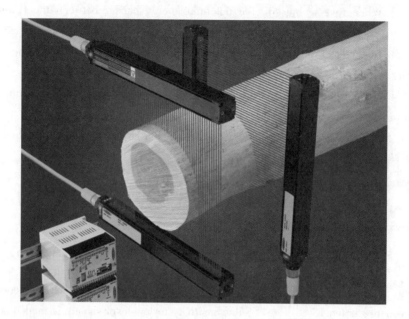

Learning Objectives

In this section, you will:

1. understand the definitions of direct detection and coherent detection.

2. understand the definitions of heterodyne detection and homodyne detection.

3. describe the block diagrams of the detection theory.

Warm-up Activity

Directions: Do you know how these applications (Fig. 4.3) work in terms of optical detection techniques?

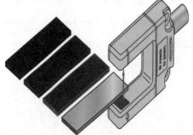

(a)

产品高度测量

(b)

Fig. 4.3   Optical detection technique applications

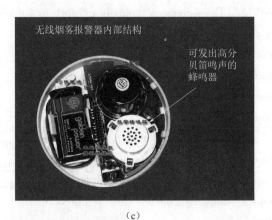

Optical Detection
Techniques
动画

(c)

(d)　　　　　　　　　　　　　　(e)

Fig. 4. 3　Optical detection technique applications（Continued）

 Text

We will review some basic concepts in optical **detection** for a better understanding of the succeeding study. The most fundamental consideration is that of the photodetector, the device that converts light into an electrical current. There are a variety of detector technologies that show high performance in the visible and infrared portions of the spectrum for use by laser radar and communications systems. For reasons of **compactness** and performance, most applications rely on modern semiconductor devices **based on** the photoelectric effect for the generation of detectable photocurrents. These photocurrents **are comprised of** photo-generated primary electrons and holes in the depletion region of the detector. Modest gains of 10 – 100 can be **achieved** in some detectors through avalanche processes, as in photomultiplier tubes（PMT）and

detection *n.* 探测，发现

compactness *n.* 紧密，简洁
base on 基于

be comprised of 由……组成

achieve *vt.* 完成，达到

avalanche photodiodes（APD）. These devices typically **exhibit** an excess noise generated by the multiplication or avalanche process, which must be taken into account in the receiver design. Still higher gains can be achieved using APD devices operating in a Geiger mode, whereby the detector is biased well beyond avalanche breakdown, resulting in very high gains（ $\sim 10^6$ ）, ultra-fast rise-times（picoseconds）, and sensitivity to single-photon events. It has been shown that in all cases, the primary photoelectron **statistics** are **identical** to the Poisson statistics of the **impinging** photon stream, while more complex statistical models are required to describe avalanche processes.

The use of photodetectors in optical receivers can be accomplished in two fundamental ways, either through direct detection or **coherent** detection. Direct detection may be considered a simple energy collection process that only requires a photodetector placed at the focal plane of a lens, followed by an electronic amplifier for signal enhancement. In comparison, coherent detection requires the presence of an **optical** local **oscillator** beam to be **mixed with** the signal beam on the photodetector surface. The coherent mixing process **imposes stringent** requirements on signal and local oscillator beam alignment in order to be efficient and can be implemented in two fundamentally different ways. If the signal and local oscillator frequencies are different and **uncorrelated**, the process is **referred to** as **heterodyne** detection, and if they are the same and correlated, as homodyne detection. Fig. 4.4 shows a **generic** optical heterodyne configuration in which the signal and local oscillator beams are generated by separate lasers of different, uncorrelated frequencies. They are combined at a beam splitter （BS）that is designed to have a reflectivity high enough to minimize signal loss but low enough to provide sufficient power for use as the local oscillator. Fig. 4.5 shows a possible **homodyne** arrangement in which a small portion of the transmit beam is used for the local oscillator, thereby satisfying the requirement for correlated frequencies.

exhibit *v.* 展出，陈列

statistics *n.* 统计学，统计表
identical *adj.* 同一的，同样的
impinge *v.* 撞击

coherent *adj.* 粘在一起的，一致的，连贯的；【物理学】相干的

optical *adj.* 眼的，视力的，光学的
oscillator *n.* 振荡器
mix with 和……混合
impose *vt.* 征税，强加，以……欺骗
stringent *adj.* 严厉的，迫切的，银根紧的
uncorrelated *adj.* 非束缚的，无关联的
refer to 查阅，提到，谈到，打听
heterodyne *adj.* ［电］外差的，外差法的
generic *adj.* ［生物］属的，类的，一般的

homodyne *adj.* ［物］零差的，零拍的

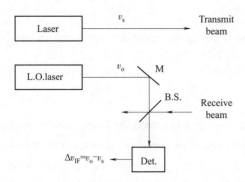

Fig. 4. 4 **A generic heterodyne detection optical configuration**

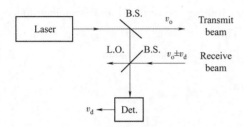

Fig. 4. 5 **A generic homodyne optical configuration with a signal Doppler shifted away from baseband by an amount v_d**

 Get to Work

I. Matching

Directions：Match the words or expressions in the left column with the Chinese equivalents in the right column.

1. photomultiplier tubes (PMT)	A. 多普勒频率
2. beam splitter (BS)	B. 电子（射）束分裂器
3. avalanche photodiodes (APD)	C. 光电倍增管
4. doppler frequency	D. 零差检波器
5. homodyne detection	E. 雪崩光电二极管

II. Understanding Checking

Directions：Mark Y (for YES) if the statement agrees with the information given in the text; N (for NO) if the statement contradicts the information given in the text.

(　　) 1. The most fundamental consideration is that of the photodetector, the device that converts an electrical current into light.

(　　) 2. The use of photodetectors in optical receivers can be accomplished in two fundamental ways, either through direct detection or coherent detection.

(　　) 3. Coherent detection requires the presence of an optical local oscillator beam to be mixed with the signal beam on the photodetector surface.

III. Passage Completion

Directions: *Fill in each of the blanks with one of the words or expressions in the box, making changes if necessary.*

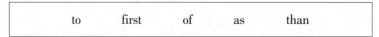

From a system point of view, there are several advantages to using homodyne detection over heterodyne detection. 1. _____, baseband homodyne detection is potentially a factor of 2 more sensitive than heterodyne detection because 2. _____ the narrower (folded) bandwidth which contains the same amount of energy in half the bandwidth of an unfolded signal, 3. _____ shown in Fig. 4. 6. Second, frequency-stabilized lasers are usually unnecessary to achieve a narrowband IF in (monostatic) homodyne systems due 4. _____ the inherent correlation of the transmitted and received frequencies, whereas frequency-stabilized and frequency-locked lasers are usually necessary in heterodyne systems. Third, a single laser system usually offers more compactness and simplicity of design 5. _____ a two-laser system.

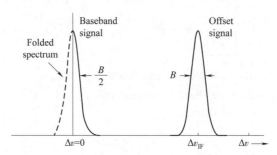

Fig. 4. 6　Typical IF spectra from a homodyne system operating at baseband and a heterodyne or offset-homodyne system operating at ΔV_{IF}. Note that the Δv_{IF} in bandwidth is for the two cases

IV. Para. Translation

Directions: *Translate the following paragraphs from English into Chinese.*

In homodyne laser radar applications, correlation between the signal and local oscillator frequencies over the round trip time-of-flight to the target τ_R is usually maintained by using a laser transmitter that has a coherence time that is longer than τ_R. In addition, the transmitter and local oscillator frequencies can be the same or different, depending on whether frequency translators are employed in the optical system. Single-laser frequency offset systems are sometimes referred to as offset-homodyne systems. Frequency offsets can also occur for signals that have been Doppler-shifted away from baseband by an amount ν_d, where $\nu_d = \pm 2|V|/\lambda$ is the Doppler frequency. The negative sign corresponds to motion away from the source so that the Doppler frequency folds over at baseband to produce a positive frequency at ν_d. These processes may still be viewed as homodyne because of the inherent correlation between the transmitted and received frequencies.

Homodyne detection has some unique properties that have proven valuable in the field of quantum optics for demonstrating the possibility of achieving photon noise levels below the quantum limit. Such noise levels FC are referred to as squeezed states, whereby the variance of amplitude or phase, but not both, is reduced below that given by Heisenberg's uncertainty principle for a coherent state. A coherent state is one in which the uncertainty principle is a minimum, otherwise known as laser light. These states are closely related to the bunching and anti-bunching statistics of the photon field.

4. 4   CCD

Learning Objectives

In this section, you will:

1. understand the structure and principle of MOS capacitor.

2. understand the operation principle of charge-coupled device (CCD) analog shift register (two phase driving).

Warm-up Activity

Do you know names of the applications based on CCD technology, as shown in Fig. 4. 7?

(a)   (b)

Fig. 4. 7   CCD technology applications

（c）

（d）　　　　　　　　　　　（e）

Fig. 4. 7　CCD technology applications（Continued）

 Text

A **CCD** is a **MOS** capacitor, the configuration of which is shown in Fig. 4. 8. An **electrode** is deposited on a **dielectric film**, formed on the semiconductor substrate.

When a voltage is applied to the electrode, a **depletion layer** is formed in the semiconductor substrate, in the region of the dielectric film. This depletion layer functions as a **pit**, in terms of energy level, for minority carriers. It is thus called a **potential well**. Applying charges to the potential well causes the charges to temporarily accumulate, thus functioning as analog memory. By varying the voltage applied to the electrode, the depth of the well can be adjusted.

A CCD analog shift register consists of a string of CCDs which are used to transfer charges. A CCD **analog shift register** is usually referred to simply as a CCD.

CCD 电荷耦合器件
MOS（metal-oxide semiconductor）［计］金属氧化物半导体
electrode *n.* 电极
dielectric film 绝缘体薄膜
depletion layer 损耗层
pit *n.* 深坑，深渊，陷阱
potential well 势垒阱

analog shift register 模拟移位寄存器

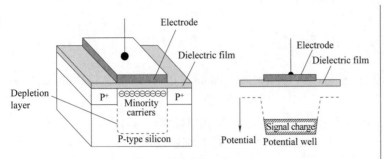

Fig. 4. 8 Basic CCD structure

This section describes the principle of operation of a CCD analog shift register, taking two-phase driving, illustrated in Fig. 4. 9, as an example.

Each MOS capacitor consists of two electrodes, under which a potential well having an uneven bottom is formed, thus determining the direction in which charges are transferred. The MOS capacitors are alternately connected to ϕ_1 and ϕ_2.

The signal charge which accumulates under the ϕ_2 electrodes, at time t_1, moves under the adjacent electrodes between times t_2 and t_3. Repeating this operation results in charge transfer.

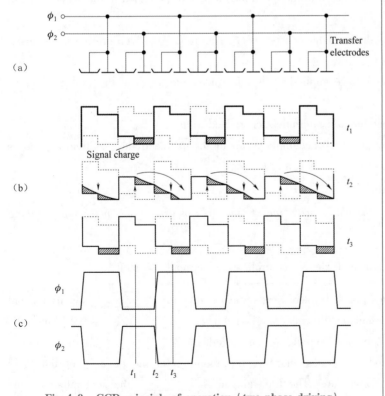

Fig. 4. 9 CCD principle of operation (two-phase driving)

Get to Work

Ⅰ. Matching

Directions：Match the words or expressions in the left column with the Chinese equivalents in the right column.

1. metal oxide semiconductor（MOS）   A. 图像分割

2. potential barrier   B. 二维图像

3. image segmentation   C. 清晰度

4. two-dimension image   D. 分辨率

5. resolution   E. 像管

6. brightness gain   F. 亮度增益

7. image element   G. 像素

8. image tube   H. 匹配

9. match   I. 势垒

10. sharpness   J. 金属氧化物半导体

Ⅱ. Understanding Checking

Directions：Mark Y（for YES）if the statement agrees with the information given in the text；N（for NO）if the statement contradicts the information given in the text.

（　　）1. When a voltage is applied to the electrode, a current is formed in the semiconductor substrate.

（　　）2. A CCD digital shift register consists of a string of CCDs which are used to transfer charges.

（　　）3. By varying the voltage applied to the electrode, the depth of the well can not be adjusted.

（　　）4. The MOS capacitors are alternately connected to ϕ_1 and ϕ_2.

（　　）5. The signal charge which accumulates under the ϕ_2 electrodes, at time t_1, moves under the adjacent electrodes between times t_2 and t_3.

Ⅲ. Passage Completion

Directions：Fill in each of the blanks with one of the words or expressions in the box, making changes if necessary.

charges	photodiode	controls	electrical	shift

The photosensitive section（Fig. 4. 10）converts optical energy to 1. _____ signals（signal charges）and temporarily accumulates the generated signal 2. _____ . Fig. 4. 10 shows the equivalent circuit of the photosensitive section, which includes a 3. _____ , MOS capacitor, and transfer gate. The photodiode generates charges, in proportion to the quantity of light and exposure time, which it then accumulates. The transfer gate 4. _____ the accumulation time （exposure time）and transfers the accumulated signal charges to the CCD analog 5. _____ register.

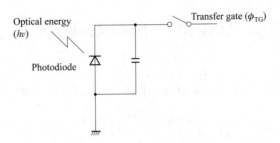

Fig. 4. 10　Equivalent circuit of the photosensitive section

IV. Translation

Directions: *Translate the following sentences from English into Chinese.*

Before color CCD linear image sensors became available commercially, color input devices were configured using one of the following methods.

1. Placing a disk having red, green, and blue color filters in front of a monochrome CCD linear image sensor, as a means of performing color separation (color filter switching).

2. Illuminating red, green, and blue light sources, in turn, inputting the produced light to a monochrome CCD linear image sensors (light source switching).

3. Using a prism to separate colors, inputting the three color components to three separate monochrome CCD linear image sensors (prism separation).

4. Mounting color filters on a monochrome CCD linear image sensor chip (on-chip color filter).

Task III   Skill Honing

4.5   Translation of Attributive Clauses
定语从句的翻译

定语从句是英语中最常用的从句之一，主要分为限制性定语从句和非限制性定语从句。其翻译方法如下。

1. 合成法。

合成法就是把定语从句译成位于名词前面的定语词组。

A semi-conductor laser is a special type of solid laser *that uses a semi-conductor*.

半导体激光器是一种使用半导体的特殊种类的固体激光器。

2. 拆分法。

如果定语从句很长或者只是对先行词进行补充说明，不太适合翻译成"的"结构的句子，翻译时可以把复合句拆开，变为一个或几个简短的并列分句。

Analogue displays have a pointer *which moves over a graduated scale*.

模拟显示器有一个指针，该指针可以在刻度盘上移动。

3. 转换法。

根据译文表达需要，可将定语从句转换为其他从句或汉语句子的其他成分，如主语、谓语、宾语等。

There is another reason *why the scientists are not in favor of revising the plan*.

为什么科学家们不赞成修改这项计划，还有另外的原因。

（定语转换为主语）

Challenge

Directions：*Translate the following sentences from English into Chinese.*

1. Mechanics is that branch of physical science which considers the motion of bodies, with rest as a special case of motion.

2. A technique that has had recent success is laser cooling.

3. The five points which the scientist stressed in his report are very important indeed.

4. It's necessary to consult the tables of technical data which are normally provided in catalogues.

5. The rod is cooled with water, which transmits the pump radiation from the flashlamp or arc lamp.

Task Ⅳ   Job Task Challenges

4.6   Career Speaking: Product Introduction

Ⅰ. Warm Up

Model

(A: Presenter   B: Mr. Black)

A: Hello, Mr. Black! I'm proud to show you the DHG smart phone. It's a new kind of phone that can make your life easier and smarter.

B: Wow, it looks *sleek* and compact. Can you talk about the camera?

A: Yes! The DHG has two cameras. They take really clear pictures even in bad light. And there's also a special lens for taking wide pictures.

B: May I have a try?

A: Sure.

B: Very convenient! How long does the battery last?

A: The battery will work all day, even if you use your phone a lot. So don't worry about it running out before bedtime!

B: That's great! What about storage space?

A: You'll have lots of space with the DHG phone! It comes with enough room to store big files like photos, music, and movies. Our phone's massive local storage greatly *surpasses* anything else in the market.

B: Terrific! Does it have a *headphone jack*?

A: Yes, it does! We want you to enjoy great sound when listening to music or watching videos on your DHG phone.

B: What else makes this phone special?

A: Our team makes sure this phone has everything you need. It has an AI processor that learns from how you use your phone. Plus, you can unlock it quickly with just your *fingerprint*.

B: It's what I wanted!

Notes

sleek	*adj.* 线条流畅的，造型优美的
surpass	*v.* 超过，优于
headphone jack	耳机插孔
fingerprint	*n.* 指纹

Ⅱ. Matching

Directions: Please choose the corresponding sentences from the box to fill in the blanks. Then, play roles with your partner.

(A: Susan   B: David)

A: Hi, David, have a look at this awesome new toy!

B: Wow! What is it?

A: It's a solar-powered car. You see the little panel on the top? _____

B: _____

A: Sure thing. Here, let me switch it on for you.

B: Look how cute it is!

A: Yeah, it's pretty cool. And no batteries needed, either.

B: So clever. _____

A: About an hour or so, depending on the brightness of the sunlight.

B: _____

A: Yeah, there are lots of designs available.

B: _____

S1: Can I take a closer look?

S2: Do they make different models?

S3: How long does it stay powered for once charged?

S4: That charges up the battery in the sun.

S5: I can't wait to get one for myself!

Ⅲ. Closed Conversation

Directions: Please fill in the blanks according to the Chinese. Then, play roles with your partners.

(A: Salesperson   B: Customer)

A: Good morning! How can I help you today?

B: I need a lamp for my desk.

A: Great, we have just the thing. It's called Smart Desk Lamp, and _____ _____ (它有不同的亮度等级，触摸感应控制，以及一个用于手机或平板计算机的充电器).

B: That sounds really good!

A: Yes, _____ (它适用于任何工作场所). And it saves electricity too.

B: Can I try it out?

A: Sure! _____ (让我告诉你怎样按住这个按钮来改变亮度).

B: Oh, that's easy. And _____ (充电器是否适用于所有设备)?

A: Yes, it works with phones, tablets, and other electronic gadgets.

B: Wow, that's awesome! But how much does it cost?

A: Right now, it's on sale for $50 instead of $70.

B: Hmm, should I buy it?

A: Definitely! _____ (这是我们最受欢迎的产品之一).

B: Okay then, I'll take it!

A: Great choice! Let me wrap it up for you.

Ⅳ. Semi-open Conversation

Directions: Please fill in the blanks with your own words. Then, play roles with your partners.

(A: Salesperson　　B: Customer)

A: Hi! How can I help you today?

B: Yes, _____

A: We have a new product called the Smart Plug. _____

B: Sounds great! Can I see it?

A: Sure thing. This is the Smart Plug. _____
You can choose from grey, white, or blue colors.

B: How much does it cost?

A: There's a special promotion now for $20.99 instead of $29.99. Would you like one in grey?

B: Yes, please!

A: Here you go. _____

B: No, thank you. You've been very helpful!

A: You're welcome. _____

Ⅴ. Live Show

Challenge 1

Directions: Lucy Green, a marketing assistant, is introducing an electronic product to Jill Jefferson, a potential customer. Please act it out with your partner or partners. The model and the useful expressions below are just for your reference.

Challenge 2

Directions: You are receiving Christina, a customer at a trade fair, and promoting your electronic products. Please act it out with your partner or partners. The model and the useful expressions below are just for your reference.

Model

(A: Rebecca　　B: Mr. Black)

A: Hi, Mr. Black. Thank you for meeting with me today to discuss our new *digital* watch.

B: Hi, Rebecca. I'm interested in learning more about the watch and I wonder what makes it different from other similar products on the market.

A: Sure thing! This watch is designed for people who have active lifestyles. It tracks steps taken, calories burned, and heart rate. But that's not all—it can also monitor sleep patterns and remind you to stand up and move around throughout the day.

B: That sounds useful. Can you show me how it works?

A: Of course. Let me demonstrate some of the features for you here on my own watch. As you can see, there's an easy-to-use touch screen *in sync with* the mobile app.

B: Wow, that looks really sleek and user-friendly.

A: Yes, and the battery life is amazing too—up to five days on a single charge. What's more, you can *customize* the style with *interchangeable* watch faces.

B: I'm definitely *intrigued*. What's the price?

A: Well, it varies depending on which model and *accessories* you choose, but we have options ranging from ＄100 to ＄250.

B: That's reasonable compared to others I've seen. I'll definitely consider purchasing one.

A: Great, I think you'll love it! Do you have any other questions or concerns?

B: No, not at this time. Thank you for your introduction.

A: My pleasure!

Notes

digital	*adj.* 数字的，数码的
in sync with	协调；与……保持一致
customize	*v.* 定制；订制
interchangeable	*adj.* 可互换的
intrigued	*adj.* 着迷的，感兴趣的
accessory	*n.* 配饰，配件

Useful Expressions

This smart phone has a high-resolution camera for stunning photos and videos.

The tablet features a large touch screen for easy navigation and viewing.

These wireless earbuds deliver crystal-clear sound with noise-cancellation technology.

The virtual assistant speaker can answer your questions, play music, and control smart home devices.

This WiFi router provides high-speed internet connectivity and advanced security features.

This 3D printer can print a variety of objects with intricate detail and precision.

The smart lock allows you to lock and unlock your door with a mobile app or voice commands.

4.7　Writing：E-mail

电子邮件的英语全称是 electronic mail，缩写为 e-mail。电子邮件具有快捷、灵活、高效等特点。用电子邮件交流是职场人必备的技能之一。电子邮件通常包括如下五个部分。

From：发件人邮箱地址。

To：收信人邮箱地址。

Date：发送日期。

Subject：邮件主题。

Message：邮件内容。

如果邮件还有附件或者发给其他人，还可以包括以下内容。

Attachment：附件，指的是与电子邮件一起发送的文字、照片、视频等。

Cc：carbon copy，抄送，指发信人将邮件发送给其他收信人，而收信人能看到列表中的其他人名单。

Bcc：blind carbon copy，密送，指发信人将邮件密送给其他收信人，而收信人看不到列表中的其他人名单。

Sample 1

From：wangmingnew@ gmail. com

To：ffzhang@ 163. com

Date：7：00 a. m. , Wednesday, June 21, 2013

Subject：Request for Information on Computers Order

Dear Miss Zhang,

I am writing to inquire about the computers order we placed with your company last week.

We would greatly appreciate it if you could provide me with some additional information regarding the order. Specifically, I would like to know about the following：

1. Price：Could you please confirm the total price of the computers? Are there any additional fees or charges that we should be aware of?

2. Shipment：When can we expect the shipment to arrive? Could you also provide me with a tracking number so that I can track the progress of the shipment?

3. Package：How will the computers be packaged? Will they be securely packed to ensure that they arrive at our office in good condition?

Thank you for your assistance in this matter.

Sincerely yours,
Wang Ming

Sample 2

From：melvinblack@ yahoo. com

To：jackbryant@ yahoo. com

Date：9：00 a. m. , Friday, June 30, 2013

Subject：Apologies for the delay in the products delivery

Dear Mr. Bryant,

I am writing to express my sincere apologies for the delay in the delivery of your ordered products. Due to **_unforeseen_** and **_adverse_** weather conditions, there has been a delay in the shipment of the goods, and they might be arriving two days later than expected.

I completely understand that this delay may cause inconvenience to you, and for that, I take full responsibility. Please know that we have already taken necessary steps to **_expedite_** the delivery process and ensure that the products reach you as soon as possible.

Once again, I apologize for any stress or inconvenience caused by the delay. If there is anything else we can do to assist you in the meantime, please do not hesitate to contact us.

Thank you for your patience and understanding.

Yours faithfully,
Melvin Black

Notes

unforeseen	*adj.* 未预见到的，无法预料的
adverse	*adj.* 不利的，有害的
expedite	*v.* 加速，促进

Useful Expressions

Thanks for your email about...

I'm writing to inquire about...

I'm interested in learning more about...

Please let me know if you need any further information.

Thanks again for all your help/time.

Could you provide me with some clarification on this issue?

Let's schedule a phone call/meeting to discuss...

We look forward to hearing from you soon.

Please see ... attached.

Enclosed you will find...

Challenge

Directions：*Assuming you are Susan（susannoble@ sina. com）, please write an email at 3：00 pm on Monday, June 5, 2023, inviting Mr. Lin（linlucky@ sina. com）to attend the product exhibition. The exhibition will be held at New World Exhibition Hall from 8：00 a. m. to 11：00 a. m. on July 10, 2023.*

Task V   Output and Evaluation

Optical Detection					
Target	Understand and describe optical detection techniques				
Requirement	Make an English presentation according to Unit 4. Draw a mind map. Produce micro-video to express professional knowledge. Add accurate Chinese and English subtitles to the video.				
Contents	**Optical Detection** Photoelectric effect Photodetector Optical detection technique				
Group：_____	Project	Name	Software	Score	Requirements
	Mind map				The logic should be sound, and the keywords used should be accurate.
	Script				The format should adhere to the specified guidelines, and it should cover all necessary contents.
	PPT				It should be consistent with the logic presented in the mind map. Clear and concise contents/Consistent formatting/Limited text/Engaging graphics and animations.
	Speaking				Clear/fluent
	Subtitle				They should be bilingual, correct and synchronized with the student's speech.
	MP4				The video should effectively convey knowledge, be accurate, and visually engaging.
Operating environment	Win7/Win8/Win9/Win10/Win11/Mobile phone				
Product features					

Optical Detection Techniques
思维导图

Optical Detection Techniques
微课

Script reference

Optical Detection

标题	专业内容（讲稿与中英字幕）	PPT 中需显示的文字	表现形式	素材
每个知识点的要点，或者小节名字	做报告时候的英文讲稿 视频中的中英字幕	需要特别提醒的文字	讲解到某个地方或某句话时，出现的图片、动画或者文字	照片，或者视频
Contents 内容	**Optical Detection** 1. Photoelectric effect 2. Photodetector 3. Optical detection techniques **光电探测技术** 1. 光电效应 2. 光电探测器 3. 光电探测技术	**Optical Detection** 1. Photoelectric effect 2. Photodetector 3. Optical detection techniques	文字	
1. Photoelectric effect 光电效应	The photoelectric effect is the name given to the observation that when light is shone onto a piece of metal, small current flows through the metal. 当光线照在一块金属上，金属中会有少量电流流过，这个现象称为光电效应。			图1　电子移动（动画图） Φ 0　0.5　1.0 mA
	The light is giving its energy to electrons in the atoms of the metal and allowing them to move around, producing the current. 光将其能量传递给金属原子中的电子，并使它们向四周运动，从而产生电流。		动态图	图1　电子移动（动画图）
2. Photodetectors 光电探测器	The most fundamental consideration is that of the photodetector, the device that converts light into an electrical current. 光电探测器能将光转变为电流。	Photodetector converts light into an electrical current.		图2　光电器件图 Photosensitive resistor　Photoelectric cell Photodiode　Photosensitive transistor

标题	专业内容（讲稿与中英字幕）	PPT 中需显示的文字	表现形式	素材
2. Photodetectors 光电探测器	Such as： Photosensitive resistor Photoelectric cell Photodiode Photosensitive transistor 例如： 光敏电阻 光电池 光电二极管 光电晶体管	Photosensitive resistor Photoelectric cell Photodiode Photosensitive transistor	依次出图显示器件	图2　光电器件图
	Optical detection techniques are widely used, for example： Fingerprint lock Scream application （to measure the crossing sectional area） rotational speed measurement illumination meter 光电探测技术应用广泛，如： 指纹锁 工件横截面积测量 转速测量仪 照度计	Applications： Fingerprint lock Scream application （to measure the crossing sectional area） Rotational speed measurement Illumination meter	图片示意实例	图3　应用举例 （图片）
3. Optical detection techniques 光电检测技术	There are a variety of detector technologies that show high performance in the visible and infrared portions of the spectrum for use by laser radar and communications systems. 在激光雷达和通信系统中使用了许多探测技术，它们在光谱的红外和可见光部分显示了很高的性能。		显示框图	图4　光电检测系统框图 （框图）
	For reasons of compactness and performance, most applications rely on modern semiconductor devicesbased on the photoelectric effect for the generation of detectable photocurrents. 为了简洁和性能方面的要求，大多数应用依赖于"基于光电效应的现代半导体器件"来产生可探测的光电流。			图4　光电检测系统框图

标题	专业内容（讲稿与中英字幕）	PPT 中需显示的文字	表现形式	素材
3. Optical detection techniques 光电检测技术	The use of photodetectors in optical receivers can be accomplished in two fundamental ways, either through direct detection or coherent detection. 在光电接收机上的光电探测器的使用由两个基本的方式组成，通过直接探测或相干探测。	Two fundamental ways：direct detection, coherent detection	根据框图进行名词解释	图5　检测技术分类框图
	Direct detection may be considered a simple energy collection process that only requires a photodetector placed at the focal plane of a lens, followed by an electronic amplifier for signal enhancement. 直接探测可以认为是一个简单的能量吸收过程，这个过程仅仅需要把一个光电探测器放在透镜的焦面上之后接一个电子放大器对信号放大增强。	Direct detection may be considered a simple energy collection process that only requires a photodetector placed at the focal plane of a lens, followed by an electronic amplifier for signal enhancement.	对照示意图讲解	图6　直接检测
	In comparison, coherent detection requires the presence of an optical local oscillator beam to be mixed with the signal beam on the photodetector surface. 相比较而言，相干探测在光电探测器的表面要有光学本地振荡器的光束和信号光束的混合存在。	In comparison, coherent detection requires the presence of an optical local oscillator beam to be mixed with the signal beam on the photodetector surface.	根据框图进行名词解释 强调coherent detection	图7　相干检测
	The coherent mixing process imposes stringent requirements on signal and local oscillator beam alignment in order to be efficient and can be implemented in two fundamentally different ways. 相干混合的过程，对信号和本机振荡光束的序列强加了严格的要求，为了能够有效地实现两种基本的不同方式。	Two fundamentally different ways.	强调coherent detection	图5　检测技术分类框图

续表

标题	专业内容（讲稿与中英字幕）	PPT中需显示的文字	表现形式	素材
	If the signal and local oscillator frequencies are different and uncorrelated, the process is referred to as heterodyne detection, and if they are the same and correlated, as homodyne detection 如果信号和本机振荡器的频率不同且无关联，则这个过程指的是外差法探测。 如果它们的频率相同且相关联，则这个过程就叫零差法探测。	If the signal and local oscillator frequencies are different and un-correlated, the process is referred to as heterodyne detection, and if they are the same and correlated, as homodyne detection.	对照框图 名词解释 强调两种方式 heterodyne detection； homodyne detection	图5　检测技术分类框图 强调两种方式
3. Optical detection techniques 光电检测技术	Fig. 8 shows a generic optical heterodyne configuration in which the signal and local oscillator beams are generated by separate lasers of different, uncorrelated frequencies. They are combined at a beam splitter that is designed to have a reflectivity high enough to minimize signal loss but low enough to provide sufficient power for use as the local oscillator. 如图8所示，由两个不同的毫无关联频率的独立激光器产生信号和本机振荡器的一般光学外差结构，它们的光束在设计成信号损失小但为本机振荡器提供极低的有效能量的分束器处结合。	A generic heterodyne detection optical configuration.	对照示意图讲解外差法探测	图8　Heterodyne detection
	Fig. 9 shows a possible homodyne arrangement in which a small portion of the transmit beam is used for the local oscillator, thereby satisfying the requirement for correlated frequencies. 如图9所示的一种零差结构，其传输光束的一小部分用于本机振荡器，因而满足频率相关的要求。	A generic homodyne optical configuration with a signal Doppler shifted away from baseband by an amount.	对照示意图讲解相干法探测	图9　Homodyne optical configuration
	The end. Thank you!	The end. Thank you!		

Task VI   Knowledge Expansion

4.8   LED Technology

Physical Principles

Like a normal diode, the LED consists of a chip of semiconducting material impregnated, or doped, with impurities to create a PN junction. As in other diodes, current flows easily from the P-side, or anode, to the N-side, or cathode, but not in the reverse direction. Charge-carriers—electrons and holes—flow into the junction from electrodes with different voltages. When an electron meets a hole, it falls into a lower energy level, and releases energy in the form of a photon.

The wavelength of the light emitted, and therefore its color, depends on the band gap energy of the materials forming the PN junction. In silicon or germanium diodes, the electrons and holes recombine by a non-radiative transition which produces no optical emission, because these are indirect band gap materials. The materials used for the LED have a direct band gap with energies corresponding to near-infrared, visible or near-ultraviolet light.

LED development began with infrared and red devices made with gallium arsenide. Advances in materials science have made possible the production of devices with ever-shorter wavelengths, producing light in a variety of colors.

LEDs are usually built on an N-type substrate, with an electrode attached to the P-type layer deposited on its surface. P-type substrates, while less common, occur as well. Many commercial LEDs, especially GaN/InGaN, also use sapphire substrate.

Colors and Materials

Zinc selenide (ZnSe) —bule;

Gallium (III) phosphide (GaP) —red, yellow, green;

Gallium arsenide phosphide (GaAsP) —red, yellow;

Aluminium gallium phosphide (AlGaP) —green;

Indium gallium nitride (InGaN) —Violet, blue;

Aluminium gallium indium phosphide (AlGaInP) —yellow, orange.

Considerations in Use

Unlike incandescent light bulbs, which illuminate regardless of the electrical polarity, LEDs will only light with correct electrical polarity. When the voltage across the PN junction is in the correct direction, a significant current flows and the device is said to be forward-biased. If the voltage is of the wrong polarity, the device is said to be reverse biased, very little current flows, and no light is emitted. LEDs can be operated on an alternating current voltage, but they will only light with positive voltage, causing the LED to turn on and off at the frequency of the AC supply.

Exercise

Directions: Translate the first paragraph into Chinese.

4. 9   Optical Storage Technology

Optical storage is a term from engineering, referring to the storage of data on an optically readable medium.

In optical-storage technology, a laser beam encodes digital data onto an optical, or laser, disk in the form of tiny pits arranged in concentric tracks on the disk's surface. A low-power laser scanner is used to "read" these pits, with variations in the intensity of reflected light from the pits being converted into electric signals.

The use of optical storage continues to grow at an incredible pace, spurred on by the flexibility and affordability the technology offers. The word "optical" in the computer industry refers to any storage method which uses a laser to store and retrieve data from media. This term includes such devices as CD-ROM, rewritable optical, WORM (standing for write-once-read-many), CD-R (short for compact disk recordable), and optical jukeboxes or autochangers.

Most of us are familiar with CD-ROM, but other terms such as rewritable optical, WORM, and CD-R may be foreign. Rewritable optical devices use media that allows data to be written repeatedly, while WORM technology writes data permanently to disk. WORM solutions were initially developed in the late 1970s but never became popular due to the lack of standardization. Today, CD-R and DVD-R (digital video disk recordable) media have replaced the non-standard devices of the 1970s. Low cost and excellent cross-drive compatibilities make it a popular WORM solution.

CD-R allows up to 650MB of data to be written permanently to a compact disk and read on low-cost CD-ROM drives. These devices are currently sold as single or stand-alone items, or as jukeboxes. Jukeboxes are storage units with built-in robotics to automate access to hundreds of pieces of media and numerous optical drives.

The continuing development of optical technology has opened up many new avenues and created limitless possibilities. Migration software now allows unused data to be moved from its original location to an optical device where it resides until it is needed. Since the only thing that touches the media is the laser, it is the most durable way to store and archive data. Optical storage solutions are also used in a wide variety of applications such as document imaging, records retention, backup systems, desktop publishing, CAD/CAM and many more.

Writable optical, when compared to other random access removable media storage solutions, is the BEST because it has the:

highest capacity available!

lowest cost per megabyte!

longest archive life of any media!

widest environmental condition tolerance!

most reliable media available!

Exercises

Directions: Fill in the blanks with the proper words or expressions given below, changing the form if necessary.

spur	storage	compatibility	CAD	archive
random access	automate	encode	built-in	imaging

1. Unfortunately, this _____ is backwards and so CD players will not play DVDs.

2. Hong Kong is the third-largest film production centre in the world, and the _____ will preserve Hong Kong's rich film heritage.

3. This is a _____ trait of human nature, so do not worry about it.

4. We should _____ the message for security reasons.

5. Primary memory is known as _____ memory and simple named memory.

6. For more than half centuries, the sciences of medical _____ grew steadily but slowly.

7. Few technologies exist to _____ the cognition process.

8. The applicants must be proficient in _____ and other office software.

9. The management accepts full responsibility for loss of goods in _____ .

10. A business tax cut is needed to _____ industrial investment.

Unit 5   Laser Processing

In this unit, you are going to learn the following contents:

Unit 5

Laser Processing

Task I Technical Principles and Device Cognition

5.1 Laser Cutting

⊙ Learning Objectives

In this module, you will:

1. understand the concept of the laser cutting.
2. learn about types of laser cutting.
3. master differences between laser cutting and mechanical cutting.
4. describe the process of laser cutting.

⊙ Warm-up Activity

Directions: Describe the pictures in Fig. 5. 1.

Words and phrases for reference: fall in love with, marriage proposal, diamond ring, come up with an idea, jeweler, laser marking.

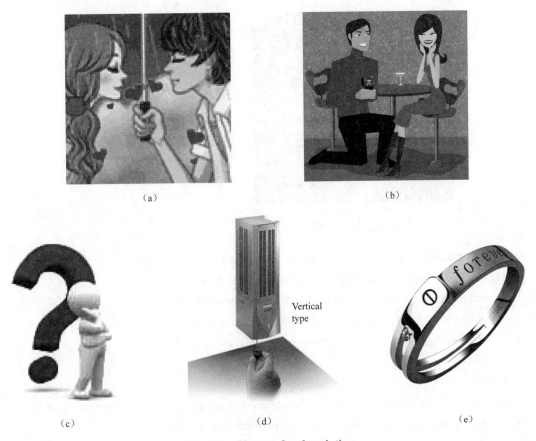

Fig. 5. 1　Pictures for description

Text

Laser cutting is a technology that uses a laser to cut materials, and is usually used in industrial manufacturing, as shown in Fig. 5. 2. Laser cutting works by **directing** the output of a high power laser, by computer, **at** the material to be cut. The material then either **melts**, burns or **vaporizes** away leaving an edge with a high quality surface finish. Industrial laser cutters are used to cut **flat-sheet materials** as well as structural and piping materials. Some 6-**axis** lasers can perform cutting operations on parts that have been pre-formed by **casting** or **machining**.

direct at 把……对准
melt v. （使）融化，（使）熔化

vaporize v. （使）蒸发
flat-sheet material 平板材料
axis n. 轴
cast v. 铸造
machine vt. 机器制造

Comparison to Mechanical Cutting

Advantages of laser cutting over mechanical cutting **vary** according to the situation, but two important factors are the lack of physical contact (since there is no cutting edge which can become **contaminated** by the material or contaminate the material), and to

vary vi. 变化，不同

contaminate v. 污染

Fig. 5. 2　Laser cutting

some extent **precision** (since there is no **wear** on the laser). There is also a reduced chance of **warping** the material that is being cut as laser systems have a small heat affected zone. Some materials are also very difficult or impossible to cut by more traditional means. One of the disadvantages of laser cutting may include the high energy required.

precision *n.* 精确，精密度，精度
wear *n.* 磨损
warp *vt.* 弄歪，使翘曲

Types

The most popular lasers for cutting materials are CO_2 and Nd：YAG, though semiconductor lasers are gaining **prominence** due to higher efficiency. Typically, there is a choice between a DC (direct current) and RF (radio frequency) powered resonator (laser generator). The fundamental choice of beam generation method can have significant impact on productivity and life cycle costs due to the differences between the two types. DC resonators have internal electrodes **encapsulated** in the cavity glass where RF resonators have external electrodes.

prominence *n.* 突出，显著

encapsulate *vt.* 装入胶囊，封装

In addition to the **power source**, the type of **gas flow** can affect performance as well. In a fast **axial flow** resonator (Fig. 5. 3), the mixture of carbon dioxide, **helium** and **nitrogen** is circulated at a high **velocity** by a turbine slab or blower. **Diffusion** cooled resonators have a static gas field that requires no **pressurization** or glassware, leading to savings on replacement of turbines and glassware.

power source 电源
gas flow 气流
axial flow 轴流
nitrogen *n.* ［化］氮

diffusion *n.* 扩散，传播
pressurization *n.* 增压，加压

Fig. 5. 3　Axial flow resonator

Process

Laser cutters usually work much like a **milling machine** for working a sheet in that the laser (equivalent to the mill) enters through the side of the sheet and cuts it through the axis of the beam. In order to be able to start cutting from somewhere else than the edge, a **pierce** is done before every cut. Piercing usually involves a high power pulsed laser beam which slowly (taking around 5 – 15 s for 0. 5 in (1 in = 0. 0254 m) thick **stainless** steel, for example) makes a hole in the material.

milling machine 铣床

pierce *v.* & *n.* 钻孔

stainless *adj.* 不锈的

 Get to Work

Ⅰ. Matching

Directions: Match the words or expressions in the left column with the Chinese equivalents in the right column.

1. high power diode laser　　　　　A. 切割质量

2. supersonic nozzle　　　　　　　B. 实时监测

3. jet flow　　　　　　　　　　　C. 光信号

4. dynamic characteristic　　　　　D. 精密切割

5. real-time monitoring　　　　　　E. 射流场

6. optical signal　　　　　　　　　F. 激光技术

7. cutting quality　　　　　　　　G. 按键面板

8. laser technique　　　　　　　　H. 动力学特性

9. keystroke　　　　　　　　　　　I. 超声速喷嘴

10. precision cutting　　　　　　　J. 大功率半导体激光器

II. Understanding Checking

Directions：*Fill in the missing parts of the short passage with no more than 3 words based on the above text.*

Laser cutting can make material melt, burn and 1. _____ leaving an edge with a high quality surface finish.

Industrial laser cutters are used to cut piping materials, structural materials and 2. _____ .

Laser cutting has many advantages over mechanical cutting, such as lack of 3. _____ , precision and less wrapping.

Even if semiconductor laser is excellent for its 4. _____ , the most popular lasers for cutting materials are CO_2 and Nd：YAG.

To start cutting from somewhere else than the edge, 5. _____ should be done before every cut.

III. Passage Completion

Directions：*Fill in each of the blanks with one of the words or expressions in the box, making changes if necessary.*

availability	offer	processing	reliable	speed
identical	due to	advantage	desirable	accomplish

When high-strength steel is concerned, the industry is increasingly turning to lasers. There are some reasons.

Firstly, in contrast to mechanical cutting, more can 1. _____ with light. The laser being an optical, non-contact tool has big 2. _____ , especially since the melting point of "usual" and high-strength steel is almost 3. _____ .

Secondly, short cycle times are especially 4. _____ . One disadvantage that laser cutting has compared to mechanical punching is its low 5. _____ . But that can be increased. Prime examples are the 3D laser 6. _____ machines of the TruLaser Cell Series 7000. They 7. _____ sensor-supported piercing on-the-fly and plasma high-speed cutting. Overall, the laser system provides increased machine power and added dynamics 8. _____ the decreased moving masses.

Thirdly, another significant factor is a high 9. _____ . The laser cutting systems, which have proven themselves in practical applications a thousand times over, are very 10. _____ .

IV. Phrase Translation

Directions：*Complete the sentences by translating into English according to the Chinese given in brackets.*

1. The input signal modulates the drive current, so the output optical signal _____ _____ （与输入电信号成比例）.

2. The trunk lines that connect central offices _____ （已被光缆所取代）.

3. Fiber optic cables _____ （对电磁噪声具有很强的抵御能力）such as

radios, motors or other nearby cables.

4. _____（光纤现在已经成为一个主要部分）of the infrastructure for a national telecommunication information highway in the US and all around the world.

5. A low-power laser scanner is used to "read" these pits, with variations in the intensity of reflected light from the pits _____（被转化为电子信号）.

V. Sentence Remaking

Directions：*Simulate the following sentence patterns according to an example provided from the text.*

1. … is … that uses … to …

E. g. Laser cutting is a technology that uses a laser to cut materials.

_____ is _____ that uses _____ to _____.

2. … is used to … as well as …

E. g. Industrial laser cutters are used to cut flat-sheet material as well as structural and piping materials.

_____ is used to _____ as well as _____.

3. … are very difficult or impossible to …

E. g. Some materials are very difficult or impossible to cut by more traditional means.

_____ are very difficult or impossible to _____.

4. In addition to …, … can … as well.

E. g. In addition to the power source, the type of gas flow can affect performance as well.

In addition to _____, _____ can _____ as well.

5. In order to …, … is done before …

E. g. In order to be able to start cutting from somewhere else than the edge, a pierce is done before every cut.

In order to _____, _____ is done before _____.

VI. Para. Translation

In 1965, the first production laser cutting machine was used to drill holes in **diamond dies**（金刚石拉模）. This machine was made by the Western Electric Engineering Research Center. In 1967, the British pioneered laser-assisted oxygen **jet cutting**（射流切割）for metals. In the early 1970s, this technology was put into production to cut **titanium**（钛）for aerospace applications. At the same time CO_2 lasers were adapted to cut non-metals, such as textiles, because, at the time, CO_2 lasers were not powerful enough to overcome the **thermal conductivity**（热导率）of metals.

VII. Mini Imitative Writing

Directions：*The following paragraph originates from the text. Please imitate it and write another one.*

Advantages of laser cutting over mechanical cutting vary according to the situation, but two important factors are the lack of physical contact, and to some extent precision. One of the disadvantages of laser cutting may include the high energy required.

5.2　Laser Marking

Learning Objectives

In this module, you will:

1. understand the concept of laser marking.

2. learn about principles of laser marking.

3. master differences between laser marking and mechanical marking.

Warm-up Activity

Directions: *Look at the principle of laser marking in Fig. 5. 4, and try to match the English phrase with Chinese meaning.*

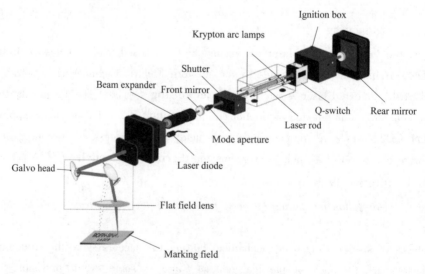

Fig. 5. 4　**Principle of laser marking**

1. ignition box	A. 激光棒
2. Q-switch	B. 激光二极管
3. krypton arc lamps	C. 点火盒
4. mode aperture	D. 调 Q 开关
5. laser rod	E. 后反射镜
6. shutter	F. 氪灯
7. rear mirror	G. 准直扩束镜
8. front mirror	H. 打标区
9. beam expander	I. 快门
10. laser diode	J. 场镜
11. galvo head	K. 选模小孔光阑
12. flat field lens	L. 前镜
13. marking field	M. 振镜

 Text

Laser marking is a non-contact **thermal** process that uses the heat generated by a laser beam to **alter** the surface of the workpiece (Fig. 5.5). The laser light must be absorbed by the material surface to generate the required heat for marking. At present, there are two types of recognized principles: by hot working and by cool working.

thermal *adj.* 热的，热量的
alter *v.* 改变

Fig. 5.5　Laser marking

Hot working uses laser beam with higher energy **density** to **irradiate** on the materials, the materials become hotter by absorbing the laser energy and with phenomena of melting, **ablation** and **evaporation**, etc.

Cool working uses high road energy **ultraviolet** photons to break materials (especially **organic** materials) or chemical bonds

density *n.* 密度
irradiate *v.* 照射
ablation *n.* 消融，切除
evaporation *n.* 蒸发（作用）
ultraviolet *adj.* 紫外线的，紫外的

organic *adj.* 有机的，组织的

153

in the mediums around, to destroy materials in the non-hot process. Cool working processing has special **significance** for laser marking, because it does not work through hot working, and will not produce hot damage or break cold **stripping** of chemical bonds, therefore, no hot **deformation** will be caused on the surface of the materials.

Compared with other marking technologies such as ink jet printing and mechanical marking, laser marking has a number of advantages, such as very high processing speeds, low operation cost (no use of **consumables**), constant high quality and **durability** of the results, avoiding **contaminations**, the ability to write very small features, and very high **flexibility** in automation.

The physical principle of laser marking is based on light amplification by simulated emission of radiation. In other words, the energy emitted by a source (e. g. arc lamp) onto an active source (e. g. Nd: YAG crystal) will be collected and concentrated between two opposing mirrors (Fig. 5. 6).

significance *n.* 意义，重要性

stripping *n.* 抽锭，脱模
deformation *n.* 变形

consumable *n.* 耗材
durability *n.* 经久，耐久力
contamination *n.* 沾污，污染，污染物
flexibility *n.* 弹性，适应性

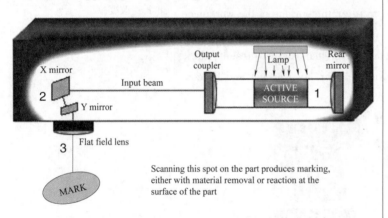

Laser Marking 动画

Fig. 5. 6　Laser marking principle

(1) The oscillation of specific light particles (photons) that move between the mirrors generates a high power laser beam. Only part of this beam passes through the output mirror (typically 10% − 20%), the rest being **reinjected** for the amplification process: the laser beam is thus created.

(2) The beam is **deflected** by two scanning mirrors, driven electronically and via software, thus enabling a very high scanning speed and **exceptional** positioning **accuracy**.

(3) The flat **field lens** then concentrates the beam energy in a very tiny spot and greatly increases the power density on the surface to be marked.

reinject *vt.* 再注入，给……作再注射
deflect *v.* （使）偏斜，（使）偏转
exceptional *adj.* 例外的，异常的
accuracy *n.* 精确性，正确度
field lens *n.* 物镜

The different markings can be classified into three categories：

The energy is **delivered** with high peak power pulses so that the material is instantaneously removed without thermal side-effects on the parts（Fig. 5. 7）.

deliver *vt.* 递送，陈述，释放
sublimation *n.* 升华，升华物

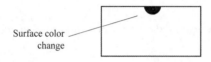

Fig. 5. 7　**Engraving with material removal（sublimation）**

The energy is delivered with lower pulses, heating the material and changing the surface appearance（Fig. 5. 8）. No material is removed.

anneal *n.* 退火，焖火

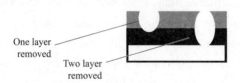

Fig. 5. 8　**Material annealing with surface color change**

On **coated** material, the contrast is created by removing the top layer, thus showing the color of the base material（Fig. 5. 9）.

Fig. 5. 9　**Marking through layer removal**

removal *n.* 移动，免职，切除
coated *adj.* 涂上一层的

Get to Work

Ⅰ. Matching

Directions：Match the words or expressions in the left column with the Chinese equivalents in the right column.

1. unit amplitude
2. tracking lidar
3. magnifying glass
4. mean dispersion
5. chemical bond
6. closed resonator
7. parallel mirror
8. zero deflection

A. 平均色散
B. 化学键
C. 平行反射镜
D. 单位振幅
E. 反射密度
F. 有机半导体
G. 跟踪激光雷达
H. 闭合共振器

9. reflection density   I. 放大镜

10. organic semiconductor   J. 零偏转，无偏转

Ⅱ. Understanding Checking

Directions：*Fill in the missing parts with no more than 3 words based on the above text.*

1. Laser marking is a _____ thermal process that uses the heat generated by a laser beam to alter the surface of the workpiece.

2. There are two types of principles：_____ and _____ .

3. Hot working makes the materials become hotter by _____ the laser energy and with the phenomena of melting, ablation and evaporation.

4. Cool working uses _____ to destroy materials in the non-hot process.

5. Laser marking has many advantages over mechanical marking, such as high speed, low cost, high quality and _____ of the result.

Ⅲ. Passage Completion

Directions：*Fill in each of the blanks with one of the words or expressions in the box, making changes if necessary.*

fiber-optics	signal	convert	modulate	distorted
copper wire	advantage	revolutionize	transmitter	role

Fiber-optic communication is a method of transmitting information from one place to another by sending light through an optical fiber. The light forms an electromagnetic carrier wave that is 1. _____ to carry information. First developed in the 1970s, fiber-optic communication systems have 2. _____ the telecommunications industry and played a major 3. _____ in the advent of the Information Age. Because of its 4. _____ over electrical transmission, the use of optical fiber has largely replaced 5. _____ communications in core networks in the developed world.

The process of communicating using 6. _____ involves the following basic steps：Creating the optical signal using a 7. _____, relaying the 8. _____ along the fiber, ensuring that the signal does not become too 9. _____ or weak, and receiving the optical signal and 10. _____ it into an electrical signal.

Ⅳ. Phrase Translation

Directions：*Complete the sentences by translating into English according to the Chinese given in brackets.*

1. The external modulator can change beam intensity _____ (不会直接影响激光的工作状态), avoiding chirp or any wavelength drift caused by pulsing the laser.

2. An ideal transmission medium would _____ (对它所传输的信号不产生影响), but any medium inevitably has some effects.

3. Both attenuation and dispersion _____ (随波长的改变而改变), so fibers have certain transmission windows.

4. Pulsed lasers which provide _____ (高冲击能) for a short period are

very effective in some laser cutting processes, particularly for piercing.

5. In optical-storage technology, a laser beam encodes digital data onto an optical disk in the form of tiny pits arranged ＿＿＿＿＿＿＿＿＿＿＿＿＿（以同心轨道方式）on the disk's surface.

V. Sentence Remaking

Directions：*Simulate the following sentence patterns according to an example provided from the text.*

1. ... has special significance for ...

E. g. Cool working processing has special significance for laser marking.

＿＿＿＿＿＿＿ has special significance for ＿＿＿＿＿＿＿＿＿＿＿＿＿＿＿＿＿＿＿＿＿.

2. Compared with ..., ...has a number of advantages.

E. g. Compared with other marking technologies, laser marking has a number of advantages.

Compared with ＿＿＿＿＿, ＿＿＿＿＿＿＿＿＿＿＿＿＿ has a number of advantages.

3. ... is based on ...

E. g. The physical principle of laser marking is based on light amplification by simulated emission of Radiation.

＿＿＿＿＿＿＿ is based on ＿＿＿＿＿＿＿＿＿＿＿＿＿＿＿＿＿＿＿＿＿.

4. ... is created by ..., thus ...

E. g. On coated material, the contrast is created by removing the top layer, thus showing the color of the base material.

＿＿＿＿＿＿＿ is created by ＿＿＿＿＿, thus ＿＿＿＿＿＿＿＿＿＿＿＿＿＿＿＿＿.

5. ... be absorbed by ...

E. g. The laser light must be absorbed by the material surface to generate the required heat for marking.

＿＿＿＿＿＿＿ be absorbed by ＿＿＿＿＿＿＿＿＿＿＿＿＿＿＿＿＿＿＿＿＿.

VI. Para. Translation

Directions：*Translate the following paragraph into Chinese.*

Markings are required on **workpieces**（工件）in all branches of industry. The attachment of numbers, texts or identity codes has become a part of the added value process. The laser has established itself as a progressive marking tools for all these tasks because it fulfills the requirements to a special degree. The **laser beam**（激光束）allows non-contact marking and is free of **wear and tear**（磨损）, making it ideal for marking any product shapes, even at hard-to-reach locations.

VII. Mini Imitative Writing

Directions：*The following paragraph originates from the text. Please imitate it and write another one.*

Compared with other marking technologies such as ink jet printing and mechanical marking, laser marking has a number of advantages, such as very high processing speeds, low operation cost, constant high quality and durability of the results, avoiding contaminations, the ability to write very small features, and very high flexibility in automation.

5.3 Laser Welding

Learning Objectives

In this module, you will:

1. understand the features of laser welding.

2. learn about two main processes: heat conduction welding and deep penetration welding.

3. master the advantages of laser welding.

Warm-up Activity

Directions: Describe the pictures in Fig. 5. 10.

(a)

(b)

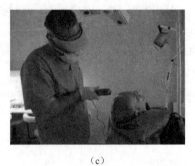
(c)

Fig. 5. 10 Pictures for description

(a) To lose one's hair; (b) Hair transplant laser; (c) Laser hair transplant

(d)　　　　　　　　(e)　　　　　　　　(f)

Fig. 5. 10　Pictures for description（Continued）
（d）New look；（e）Job interview；（f）Success

Text

The laser beam welding（Fig. 5. 11）is mainly used for joining **components** that need to be joined with high welding speeds, thin and small weld **seams** and low thermal distortion. The high welding speeds, an excellent automatic operation and the possibility to control the quality online during the process make the laser welding a common joining method in the modern industrial production.

component n. 成分

seam n. 接缝，线缝

Fig. 5. 11　Laser beam welding

In laser welding, one must **distinguish** between two main processes, heat **conduction** welding and deep **penetration** welding.

In heat conduction welding（Fig. 5. 12（a）），the materials to be joined are melted by **absorption** of the laser beam at material surface—the **solidified** melt joins the materials. Welding penetration depths in this context are typically below 2 mm.

Deep penetration welding（Fig. 5. 12（b）），which starts at energy densities of **approx.** 106 W/cm^2, is based on the creation of a vapor **capillary** inside the material by local heating to the

distinguish v. 区别，辨别
conduction n. 传导
penetration n. 穿过，渗透，突破
absorption n. 吸收
solidify v.（使）凝固，（使）团结
approx. abbr. approximate adj. 近似的，大约的
capillary n. 毛细管；adj. 毛状的，毛细作用的

evaporation temperature. The resulting vapor pressure inside the material creates a capillary approx. 1.5 times the diameter of the **focal spot** of the laser beam, which is moved through the material by the device, following the **contour** to be welded. The **hydrostatic** pressure, the surface tension of the melt, and the vapor pressure inside the capillary reach a balance, preventing the capillary (often referred to as the "**keyhole**") from **collapsing**. Multiple reflection inside the keyhole guides the incident laser beam deep into the material. Today, given sufficient laser power, weld depths of up to 25 mm (steel) can be achieved.

focal spot 焦斑（点）
contour *n.* 轮廓，周线，等高线
hydrostatic *adj.* 静水力学的，流体静力学的
keyhole *n.* 锁眼
collapse *vi.* 倒塌，崩溃，瓦解

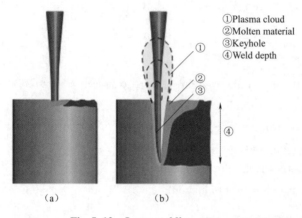

①Plasma cloud
②Molten material
③Keyhole
④Weld depth

（a）　　　　　（b）

Fig. 5.12　Laser welding process
（a）Heat conduction welding；（b）Deep penetration welding

Compared to **conventional** welding methods, laser welding offers the **diverse** advantages: no tool wear, contact-free processing, diverse materials, and high welding seam quality. **Provided** the two parts to be joined is good (i.e. a small gap relative to the thickness of material), no additional material need to be added, such as **welding rod**. There are occasions however when it is **desirable** to add material, for example:

conventional *adj.* 惯例的，传统的
diverse *adj.* 不同的，变化多的
provided *conj.* 倘若

welding rod 焊条
desirable *adj.* 值得要的，合意的

When **dissimilar** materials cannot be welded together to **yield** a strong joint, **metallurgy** can point to a suitable **intermediate** material which can be added to the weld in rod form to make a **crackfree** joint possible. In tool repair, **worn** surfaces can be built up so that tools can be **refurbished**. In jewellery repair, it is often necessary to add material to effect repairs such as **reprocessing** jewel mounts or **resizing** rings. Laser welding with fine wire of the same material as the original item makes this possible.

dissimilar *adj.* 不同的，相异的
yield *v.* 出产，生长，生产
metallurgy *n.* 冶金，冶金术
intermediate *adj.* 中间的；*n.* 媒介
crackfree *adj.* 无裂缝的
worn *adj.* 用旧的，疲倦的
refurbish *vt.* 再磨光，刷新
reprocess *v.* 再生，再加工
resize *v.* 调整大小

Get to Work

Ⅰ. Matching

Directions：Match the words or expressions in the left column with the Chinese equivalents in the right column.

1. photoreader
2. photojunction diode
3. adaptive filter
4. three-dimensional laser processing
5. all-purpose instrument
6. image sensor
7. discharge tube
8. passive imaging system
9. absorption coefficient
10. optical memory

A. 吸收系数
B. 自适应滤光片
C. 三维激光加工
D. 光学存储器
E. 光电读出器
F. 被动成像系统
G. 图像传感器
H. 通用仪器
I. 光电二极管
J. 放电管

Ⅱ. Understanding Checking

Directions：mark Y（for YES）if the statement agrees with the information given in the text；N（for NO）if the statement contradicts the information given in the text.

(　　) 1. Laser welding is widely used for joining components because of its high welding speeds and low thermal distortion.

(　　) 2. There are two main processes of laser welding, that is, heat welding and cool welding.

(　　) 3. Deep penetration welding depth can range from 2 mm to 25 mm.

(　　) 4. If the two parts to be joined have a small gap, welding rod need be added.

(　　) 5. When dissimilar materials cannot be welded together, a suitable intermediate material can be added to the weld to make a crackfree joint possible.

Ⅲ. Passage Completion

Directions：Fill in each of the blanks with one of the words or expressions in the box, making changes if necessary.

announce	abandon	replace	all-around	single standard
as of	support	while	storage format	higher-quality

The beginnings of the DVD format may be traced back to the 1990s, where two optical 1. ＿＿＿＿＿ were in the works. Philips and Sony had the multimedia compact disk（MMCD）, 2. ＿＿＿＿＿ numerous companies including Toshiba, Time-Warner, Matsushita Electric, Mitsubishi Electric, Pioneer, Thomson, and JVC 3. ＿＿＿＿＿ the super density disk（SDD）. However, then-president of IBM Lou Gerstner moved for a 4. ＿＿＿＿＿ and a unification of two camps. The MMCD 5. ＿＿＿＿＿ and the SD format took on a number of revisions. In 1995, the

DVD specification version 1. 5 6. _____ . By 1997, the DVD Forum 7. _____ the DVD Consortium (联盟) and was opened to all companies.

8. _____ today there is no official meaning behind the letters DVD. The term "Digital Versatile Disk" came first, as the primary purpose of the DVD was to fully realize a 9. _____ home theater experience. Some members of the DVD Forum, however, believe "Digital Versatile Disk" is more appropriate to denote 10. _____ use and not just video.

IV. Phrase Translation

Directions：*Complete the sentences by translating into English according to the Chinese given in brackets.*

1. With sufficiently high energy density, the laser beam heats the material to such a point, that it is _____ (不仅在表面熔化，而且蒸发).

2. In military and defense, _____ (快速传递大量信息) is the impetus behind a wide range of retrofit and new fiber optic programs.

3. It was Bell himself who _____ (发明了最早的光波通信设备) in 1880.

4. This location, called a node, would provide the optical receiver that _____ _____ (将光脉冲转化为电信号).

5. In a fiber, the two principal limiting effects are attenuation of the signal strength and _____ _____ (脉冲的色散).

V. Sentence Remaking

Directions：*simulate the following sentence patterns according to an example provided from the text.*

1. … make … a … in …

E. g. The high welding speeds make the laser welding a common joining methodin the modern industrial production.

_____ make _____ a _____ in _____ .

2. …, which …, is based on …

E. g. Deep penetration welding, which starts at energy densities of approx. 106/cm^2, is based on the creation of a vapor capillary inside the material.

_____ , which _____ ,
is based on _____ .

3. Given …, … can be achieved.

E. g. Given sufficient laser power, weld depths of up to 25 mm (steel) can be achieved.

Given _____ , _____ can be achieved.

4. Provided …, no … need …

E. g. Provided the two parts to be joined is good (i. e. a small gap relative to the thickness of material), no additional material need to be added.

Provided _____ , no _____

need _____ .

5. … make … possible.

E. g. Laser welding with fine wire of the same material as the original item makes this possible.

_____ make _____ possible.

VI. Para. Translation

Directions：Translate the following paragraph into Chinese.

The specific features of manual welding systems have revolutionized manufacturing and repair work in the jewelry industry. Work pieces are manually positioned in a working chamber and viewed through a **microscope**（显微镜）with **crosshair**（十字瞄准线）. They are permanently spot welded making expensive **clamps**（夹钳）or time-consuming wire binding unnecessary. Precious metals and **alloys**（合金）can be welded without any **filler**（填充物）or **soldering material**（焊接材料）. Laser weld seam is extremely robust and requires only little post processing. **Impurities**（杂质）of potentially **toxic**（有毒的）alloys are avoided.

VII. Mini Imitative Writing

Directions：The following paragraph originates from the text. Please imitate it and write another one. Define a device or technique related to laser.

The laser beam welding is mainly used for joining components that need to be joined with high welding speeds, thin and small weld seams and low thermal distortion. The high welding speeds, an excellent automatic operation and the possibility to control the quality online during the process make the laser welding a common joining method in the modern industrial production.

5.4 Laser Perforating

Learning Objectives

In this module, you will:

1. understand the purpose of laser perforating.

2. learn about applications of laser perforating.

3. know process of how to make "easy open" packaging.

Warm-up Activity

Directions: Please write an individual signature (个性签名) for each of functions of laser perforation shown in Fig. 5. 13. You may discuss with your partner with help of Fig. 5. 13 (a) as an example. Some of you will be invited to show your created individual signatures to the whole class.

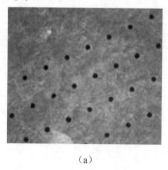

(a)   (b)

Individual signature: I'm the mascot   Individual signature: _____
of smokers!   _____
                                         _____.

Fig. 5. 13　Function of laser perforation

(a) Cigarette tipping paper; (b) Atmosphere packaging

（e）

（d）

Individual signature： _____

_____

_____ .

Individual signature： _____

_____

_____ .

Fig. 5. 13　Function of laser perforation（Continued）

（c）Easy open packaging；（d）Anti-forgery passport

 Text

Lasers are used to drill tiny holes at high speed in webs of thin paper and plastic （Fig. 5. 14） . The purpose of these **perforations** is generally to create weaknesses to allow packages to open in a **predefined** way （"Easy Open"）or to make the material **porous** to allow air or steam to move through the barrier in a controlled way.

perforation *n.* 穿孔

predefine *vt.* 预先确定

porous *adj.* 多孔的，渗水的

Fig. 5. 14　Multiple laser heads perforate plastic and paper in-line

Application examples are as below.

Perforation of Cigarette Tipping Paper

By producing **an array of** holes in the range of 50 – 100 μm in tipping paper, cigarette manufacturers reduce the **tar inhaled** by the smoker as cool air is drawn into the filter, which encourages the tar vapour to **condense** inside the filter, rather than being inhaled into the **lungs**.

A kilowatt CO_2 laser is split into multiple beams and delivered to the paper by means of **stationary** microfocusing heads. In the

tipping paper 接装纸

an array of 一排，一群，一批

tar *n.* 焦油，柏油

inhale *vt.* 吸入

condense *v.* （使）浓缩，精简

lung *n.* 肺，呼吸器

stationary *adj.* 固定的

most rapid winding systems, such systems are capable of perforating at approaching 500,000 holes/s!

Laser Perforation of Modified Atmosphere Packaging

As cut fruit and vegetables continues to **respire** within plastic packaging, the **shelf life** of produce can be extended by making the packaging "**breathable**" —i. e. **permeable** to oxygen and carbon dioxide. Laser drilled holes reliably create the correct size and density of holes to **facilitate** this without **drying out** the produce.

"Easy Open" Packaging

Lasers (Fig. 5.15) can **scribe** or perforate material which will subsequently be used to package candies, snacks, coffee, and drinks to make the product easier to access. Tear lines can be created by scribing or making a **linear** perforation to weaken the packaging where it is required. The CO_2 laser is well absorbed by organic materials but not by foil barriers used in peanut and crisp packs meaning that the contents remain **sealed** from atmosphere.

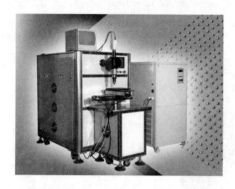

Fig. 5.15 Laser perforation of packaging

Similarly, entry holes for **drinking straws** and exit areas for **blister** packs can be made so much easier to use by selective package weakening. Creating such structure on a web of material whilst, it moves between **rollers** which can require high speed movement of the laser spot across the web. Generally, these applications therefore require "**galvo**" optics which deflect the laser at a very high speed using rotating pairs of mirrors.

Anti-forgery Ticket and Passport Perforation

Passports can have the number cut through all the pages by **steering** the laser to create **alphanumerics** made from a **matrix** of circles. World cup tickets have been finished with **intricate tear-off** perforations which can be very difficult for **forgers** to **emulate**.

modified *adj.* 改良的，改进的，修正的

respire *v.* 呼吸

shelf life （包装食品的）货架期，保存期限

breathable *adj.* 通气的，可渗透的

permeable *adj.* 有浸透性的，能透过的

facilitate *vt.* 使便利，促进

dry out 干透，使干

scribe *vt.* 用划线器划

linear *adj.* 线的，直线的，线性的

sealed *adj.* 未知的，密封的

drinking straw 吸管

blister *n.* 水泡

roller *n.* 滚筒，辊子

galvo *n.* 检流计

anti-forgery *n.* 防伪

steer *v.* 控制，引导，驾驶

alphanumeric *adj.* 文字数字的

matrix *n.* 矩阵

intricate *adj.* 复杂的，错综的

tear-off *n.* 可按虚线撕下的纸

forger *n.* 伪造者

emulate *vt.* 仿效

Get to Work

Ⅰ. Matching

Directions：Match the words or expressions in the left column with the Chinese equivalents in the right column.

1. electro-optics	A. 射束发散性
2. light-wave	B. 红外线谱
3. trunk line	C. 电光学
4. milling machine	D. 输出耦合
5. beam divergence	E. 横模，横向模态
6. transverse mode	F. 光波
7. thermal lensing	G. 干线，中继线
8. infrared spectrum	H. 掺杂浓度
9. output coupling	I. 铣床
10. doping concentration	J. 热透镜效应

Ⅱ. Understanding Checking

Directions：Choose the best answer for each statement below according to the text.

1. The purpose of laser perforations is generally to create _____ to allow package easy open.

    A. strength   B. energy source   C. weakness   D. barrier

2. Laser perforation can make fruit and vegetables continue to _____ within plastic packaging.

    A. grow   B. extend   C. respire   D. breathe

3. "Easy open" packaging is often used to package the following except _____ .

    A. snacks   B. sweetmeat   C. juice   D. cigarette tipping paper

4. Galvo optics _____ the laser at very high speed using rotating pairs of mirrors.

    A. radiate   B. deflect   C. convey   D. distort

5. The word "alphanumerics" in the second line of last paragraph means _____ .

    A. letter and number   B. alphabet   C. picture   D. figure

Ⅲ. Passage Completion

Directions：Fill in each of the blanks with one of the words or expressions in the box, making changes if necessary.

develop	prospect	appropriate	combine with	efficiency
non-metallic	widely	promising	extremely	suitable

Laser is the new light sources in 1960s. It is 1. _____ used due to the lasers' good direction, high brightness. Laser processing is one of the most 2. _____ fields in laser application, now more than 20 kinds of laser processing technology have been 3. _____ .

Laser's space and time control is very good, there are very large spaces for the material of the object, shapes, sizes and processing conditions, especially 4. _____ for automation processing. Laser processing system 5. _____ computer numerical control technology can form a high efficient automatic processing equipment which has become a key technology for the enterprise's production. It has opened up a broad 6. _____ for high quality, high 7. _____ and low cost manufacturing.

Both hot processing and cold processing can be used in metallic and 8. _____ materials, like cutting, drilling, marking. Hot processing is 9. _____ beneficial for metal materials welding, surface treatment, alloy production, cutting. Cold processing is 10. _____ for light chemical deposition, rapid laser prototyping technology, laser etching, mixed dyeing and oxidation.

IV. Phrase Translation

Directions: Complete the sentences by translating into English according to the Chinese given in brackets.

1. The advent of lasers in 1960 and low-loss optical fiber in 1970 _____
_____ (消除了所有这些障碍).

2. In recent years, it has become apparent that _____ (光纤正一步步替代铜线) as an appropriate means of communication signal transmission.

3. Since _____ (唯一接触媒介的物质) is the laser, it is the most durable way to store and archive data.

4. _____ (光学技术的持续发展) has opened up many new avenues and created limitless possibilities.

5. Juke boxes are storage units _____ (有内置机器人技术) to automate access to hundreds of pieces of media and numerous optical drives.

V. Sentence Remaking

Directions: simulate the following sentence patterns according to an example provided from the text.

1. The purpose of ... is generally to ...

E. g. The purpose of these perforations is generally to create weaknesses to allow packages to open in a predefined way.

The purpose of _____ is generally to
_____ .

2. ... is ... but not ...

E. g. The CO_2 laser is well absorbed by organic materials but not by foil barriers used in peanut and crisp packs.

_____ is _____
but not _____ .

3. As ..., ... can be extended.

E. g. As cut fruit and vegetables continues to respire within plastic packaging, the shelf life of produce can be extended.

As _____ , _____ can

be extended.

4. Similarly, ... can be made ...

E. g. Similarly, entry holes for drinking straws and exit areas for blister packs can be made so much easier to use by selective package weakening.

Similarly, _____ can be made _____ .

5. ... have been finished with ...

E. g. World cup tickets have been finished with intricate tear-off perforations.

_____ have been finished with _____ .

VI. Para. Translation

Directions: Translate the following paragraphs into Chinese.

1. **Cigarette-tip** （烟嘴） paper is perforated mainly to reduce the tar and nicotine content of cigarettes. Laser perforation ensures the definite adherence to the **threshold values** （界限值）.

2. The laser perforation of packaging films is used to prolong the freshness and quality of **perishable** （易腐坏的） food. Micro perforation enhances the shelf life of vegetables etc. by exchanging oxygen through micro holes in the range of $60 - 100$ μm.

3. The individual perforation of identification documents or VIP entrance cards with CO_2 lasers guarantees to gain maximum protection against forgery.

VII. Mini Imitative Writing

Directions: The following paragraph originates from the text. Please imitate it and write another one. Describe a device's function.

Lasers can scribe or perforate material which will subsequently be used to package candies, snacks, coffee, and drinks to make the product easier to access. Tear lines can be created by scribing or making a linear perforation to weaken the packaging where it is required.

Task Ⅱ Honing Skills

5.5 Translation of Adverbial Clauses
状语从句的翻译

英语的状语从句一般由从属连接词和起连词作用的词组引导。英语状语从句包括时间、目的、地点、原因、条件、让步、结果、比较等从句，可采用顺译法、换序法、分译法、合译法、转换法，及几种方法同时采用的并用法。

1. 顺译法。

把状语从句直接汉译，并与原文位置相同。

This process continues *till an equilibrium is established*.

这个过程持续下去直到达到平衡。

2. 换序法。

依据汉语语言习惯或表意需要，状语从句汉译后，与原文相比，位置发生变化，或前移、或后移、或中置。

Usually the capacitor is made up of plates of large area *so that large electrical charges may be stored*.

为了能储存大量电荷，电容器通常用大面积的金属板来制造。

3. 分译法。

有的状语从句非常长，且结构复杂。采用分译法能清楚表达长句的意思。

Albert Einstein's paper—On the Quantum Theory of Radiation in 1917 was considered as a great event, *because it laid the foundation for the invention of the laser and its predecessor, the maser, in a ground-breaking rederivation of Max Planck's law of radiation based on the concepts of probability coefficients (later to be termed "Einstein coefficients") for the absorption, spontaneous and stimulated emission*.

阿尔伯特·爱因斯坦的论文《论辐射的量子性》在 1917 年发表被认为是重大事件，因为该论文为激光器及其先驱——微波激射器的发明奠定了基础。这项成果源于对马克斯·普朗克辐射定律创造性的再次运用。而马克斯·普朗克辐射定律基于概率系数理论（该系数后来被称为"爱因斯坦系数"），用于研究受激吸收、自发辐射和受激辐射。

4. 合译法。

含有状语从句的复合句在汉译时可省去主句的主语或从句的主语，将两个句子组合成一个单句。

The iron should be stricken while it is hot.

趁热打铁。

5. 转换法。

译文根据需要，可将一类状语从句译为另一类从句或词组，如条件状语从句译为时间状语从句。

If all the steps are taken in accordance with the procedures, the system will be recovered.

当依程序进行所有步骤，系统将恢复。(转译为时间状语从句)

6. 并用法。

由于表意需要，有些状语从句在汉译时需要将以上几种方法并用。

Shall the atoms influence process of change if the atoms move?

原子运动是否会影响变化的进程？

译文采用换序法，调整主句与状语从句的语序，将状语从句提前，主句放后；同时，采用了合译法，将主句和从句合成一个单句。

Challenge

Directions：*Translate the following sentences from English into Chinese.*

1. Whenever a wave moves out from a source in uniform medium, it travels in straight lines.

2. Where there is the Internet, there is chatting.

3. The path is not completed till wires are connected.

4. Current stops flowing as soon as we break the circuit.

5. Chemical reaction occurs if possible.

6. Information becomes jammed in a bottleneck at the points where conversion to, or from, electronic signals is taking place.

7. As the sphere becomes larger, the waves become weaker.

Task Ⅲ Challenging Tasks

5.6 Career Speaking：After Sales Service

Ⅰ. Warm Up

Model

（A：customer B：receptionist）

A：I bought an ***air-conditioner*** at a supermarket. But it had only been used for one week before it couldn't work.

B：I suggest you ask that supermarket to repair or replace it. They are responsible for providing after-sales service.

A：Yes，I did. But the manager of the supermarket told me they had sold out all the air-conditioners with that model. What's worse，they were short of some ***spare parts***. So they asked me to turn to you for help.

B：Could you tell me what your model number is?

A：PIU-9254.

B：All right. Our worker will be sent to your place tomorrow morning if it's convenient to you.

A：Oh，unfortunately，I'll go abroad on business tomorrow. How about this afternoon? I've asked for leave to stay at home.

B：Let me check whether there is someone available. You know，most of our workers have gone out to deliver goods.

A：I'm sorry to have given you so much trouble.

B：It doesn't matter，sir. Well，Paul is free. He'll get there two hours later. Please wait for him at home.

A：OK，thank you.

B：You're welcome. Is there anything else I can help you?

A：No，I'm satisfied with your service.

B：Thanks for your support.

Notes

| air-conditioner | *n.* 空调 |
| spare part | 备件 |

Ⅱ. Matching

Directions：Please choose the corresponding sentences from the box to fill in the blanks. Then，play roles with your partner.

（A：assistant manager B：customer）

A：Hello，this is Customer Service Department of Grace Electrical Equipment Company. ____

B：Hello, I ordered 3,000 TV sets from your company last month, but I find 20 of them can't work.

A：I'm sorry to hear that. _____

B：Yes, we've already sent them.

A： _____

B：All right. How long will it take?

A：_____ I feel sorry again.

S1：If the responsibility actually lies in us, we will send you a replacement right away and compensate your loss.

S2：Would you like to ship back the faulty ones in case that we should make a through investigation?

S3：May I help you?

S4：Almost one week after the arrival of the goods.

Ⅲ. Closed Conversation

Directions：Please fill in the blanks according to the Chinese. Then play roles with your partner.

（A：customer　　B：secretary）

A：I'm satisfied with your product, but _____?
（但是我想知道你们的售后服务怎么样?）

B：You may rest assured. _____. （我们关注产品的声誉，并且会尽力使顾客愉快的。）

A：If there's something wrong with the product, will it be replaced or repaired for free?

B：_____, （我们只在一个月内包换,）but we offer free repair within three years from the purchase date as long as you keep the receipt and the warranty card.

A：_____? （操作时遇到问题怎么办?）

B：You may contact us by phone, letter or email. _____
_____. （我们的专业人员会负责解决问题的。）

A：That's quite good. Thank you.

Ⅳ. Semi-open Conversation

Directions：Please fill in the blanks with your own words. Then play roles with your partner.

（A：customer　　B：sales manager）

B：May I help you, Madam?

A：_____.

B：What's the matter?

A：_____.

B：Is that so? So far we haven't had any complaint of this kind.

A：I've brought some of them. _____.

B：Oh，I'm terribly sorry. _____.

A：How do you deal with the problem?

B：Please return the wrong pieces for our account. We'll deliver a replacement within two days.

_____.

V. Live Show

Challenge 1

Directions：A customer complained about the notebook he bought and asked to repair or replace it. You are in charge of dealing with the matter. Please act it out with your partner or partners. The model and the useful expressions below are just for your reference.

Challenge 2

Directions：You call your customers to collect their response to the products and solve the problems. Please act it out with your partner or partners. The model and the useful expressions below are just for your reference.

Model

（A：Lucy，salesgirl   B：Mr. Black，customer）

A：Hello，is that Mr. Black please?

B：Yes，who's that speaking?

A：This is Lucy speaking，from Customer Service Department of Joy Trade Company. I'm calling to ask if the delivery of your ***projector*** went OK.

B：Your delivery is very efficient. And the workers were very patient to tell me how to operate and maintain it.

A：It sounds good. Is there anything wrong with the projector or do you have any questions about it?

B：No，it is working very well. The ***manual*** is helpful to solve my problems.

A：You can contact us at all times. After receiving your trouble call，we'll visit you in 24 hours.

B：Thank you. By the way，what is the customer service hotline?

A：800 – 8452 – 3986.

B：800 – 8452 – 3986. I've got it. Can I update the product?

A：Yes. We'll update it for free within one year. And after that you can replace it with a new one at 30% discount.

B：Really? That's good news.

A：Mr. Black，please make comments on our service to help us improve our work.

B：Quite good. I'm very satisfied. I'll ***recommend*** your company to my colleagues and choose more products.

A：Thanks for your trust and encouragement. We won't ***fail*** you.

Notes

projector   *n.* 投影仪

manual	*n.* 说明书
recommend	*v.* 推荐
fail	*v.* 使失望

Useful Expressions

1. Our customer satisfaction scores are among the best in the industry.

2. For any questions, please call our customer service hotline at 400 – 6512 – 3971.

3. We'll deliver and install it for you.

4. When do you want to deliver it?

5. Do you want the product repaired or replaced?

6. If there is any fault, you can replace it.

7. The factory offers a 4 year's warranty for this computer.

8. Please send it to our address for repair.

9. Because we're not careful about quality control, there are some faulty materials.

10. Can you tell me the brand and model?

11. Goods purchased in this plant can be changed in 7 days.

5.7　Writing: Letter of Apology

在工作和生活中，由于疏忽或过失做错了事，给他人造成了麻烦、伤害或损失，要立即写信给对方赔礼道歉。道歉信通常包括三部分内容：1. 表示道歉的缘由；2. 提出补救的措施；3. 再次道歉，请求谅解。道歉信的语言要真诚恳切，理由要真实可信，措施要切实可行，切忌掩盖事实、强词夺理。

Sample 1

Dear Mr. Clark,

I am writing this letter to apologize to you for my failing to attend to day's meeting.

My wife suddenly had a high fever this morning, and I had to take her to the hospital and stayed there to take care of her. Until my sister arrived, I left and drove back in a hurry. I had thought I could be in time for the meeting but I still missed it. I know the importance of the meeting so I'll ask Sally to let me read the ***minutes*** and then I'll write a report to you.

Once again, I apologize for the inconvenience that my absence may cause.

Cordially yours,
Ben

Note

| minutes | *n.* 会议记录 |

Sample 2

Dear Mr. White,

I would like to give you my apology for the delay in ***delivery***. This is due to a thunderstorm that destroyed our plant. Now we're busy ***work***ing ***in three shifts*** so as to speed up the shipment. We'll be able to complete the delivery by the end of August. In order to ***compensate*** your possible loss, we'll give you another 2% *discount*.

I sincerely hope that you can understand our position and accept my apology.

Yours faithfully,
Jerry Morris

Notes

delivery	*n.* 交付，交货
work in three shifts	三班倒
compensate	*v.* 补偿
discount	*n.* 折扣

Useful Expressions

I would like to give you my apology for...

Please allow me to say sorry again.

We would appreciate your understanding and ask you to accept our apology.

Please accept my sincere apology for...

It's due to my negligence, for which I'm exceedingly sorry.

Challenge

Directions：You haven't finished the financial report your manager asked you to do, so you have to write a letter of apology to explain reasons and ask for understanding.

Task Ⅵ   Outputting and Evaluating

<table>
<tr><td colspan="6" align="center">Laser Marking</td></tr>
<tr><td>Target</td><td colspan="5">Understand and describe a basic circuit</td></tr>
<tr><td>Requirement</td><td colspan="5">Make an English presentation according to lesson 5. 2.
Draw a mind map.
Produce Micro-video to express professional knowledge.
Add accurate Chinese and English subtitles to the video.</td></tr>
<tr><td>Contents</td><td colspan="5">Laser Marking
Hot working and cool working
Advantages.
The physical principle of laser marking
Three categories.</td></tr>
<tr><td rowspan="7">Group： _____</td><td>Project</td><td>Name</td><td>Software</td><td>Score</td><td>Requirements</td></tr>
<tr><td>Mind map</td><td></td><td></td><td></td><td>The logic should be sound, and the keywords used should be accurate.</td></tr>
<tr><td>Script</td><td></td><td></td><td></td><td>The format should adhere to the specified guidelines, and it should cover all necessary contents.</td></tr>
<tr><td>PPT</td><td></td><td></td><td></td><td>It should be consistent with the logic presented in the mind map.
Clear and concise contents/Consistent formatting/Limited text/Engaging graphics and animations.</td></tr>
<tr><td>Speaking</td><td></td><td></td><td></td><td>Clear/fluent</td></tr>
<tr><td>Subtitle</td><td></td><td></td><td></td><td>They should be bilingual, correct and synchronized with the student's speech.</td></tr>
<tr><td>MP4</td><td></td><td></td><td></td><td>The video should effectively convey knowledge, be accurate, and visually engaging.</td></tr>
<tr><td>Operating environment</td><td colspan="5" align="center">Win7/Win8/Win9/Win10/Win11/Mobile phone</td></tr>
<tr><td>Product features</td><td colspan="5"></td></tr>
</table>

Laser Marking
思维导图

Laser Marking
微课

Script Reference

标题	专业内容（讲稿与中英字幕）	PPT 中需显示的文字	表现形式	素材
每个知识点的要点，或者小节名字	做报告时候的英文讲稿 视频中的中英字幕	需要特别提醒的文字	讲解到某个地方或某句话时，出现的图片、动画或者文字	照片，或者视频
	Laser marking is a non-contact thermal process that uses the heat generated by a laser beam to alter the surface of the workpiece. 激光打标是一个非接触性的热反应过程。在这个过程中通过激光束所产生的热能来改变工件的表面。	Laser marking is a non-contact thermal process that uses the heat generated by a laser beam to alter the surface of the work-piece.		图1　激光打标，效果图
Laser marking 激光打标	**Laser Marking** Contents： 1. Hot working and cool working 2. Advantages 3. The physical principle 4. Three categories 主要内容 1. 热加工和冷加工 2. 优点 3. 物理原理 4. 三类	**Laser Marking** Contents： 1. Hot working and cool working 2. Advantages 3. The physical principle 4. Three categories		
1. Hot working and cool working 热加工和冷加工	At the present, there are two types of recognized principles：by hot working and by cool working. 目前有两种公认的打标原理：加热处理和冷却处理。 Hot working uses laser beam with higher energy density to irradiate on the materials, the materials become hotter by absorbing the laser energy and with phenomena of melting, ablation and evaporation, etc. 热加工使用高能量密度的激光束照射在材料表面，材料通过吸收激光能量变热，从而发生熔化，切除和蒸发等现象。	At the present, there are two types of recognized principles：by hot working and by cool working. Hot working uses laser beam with higher energy density to irradiate on the materials, the materials become hotter by absorbing the laser energy and with phenomena of melting, ablation and evaporation, etc.	文字	

标题	专业内容（讲稿与中英字幕）	PPT 中需显示的文字	表现形式	素材
1. Hot working and cool working 热加工和冷加工	Cool working uses high road energy ultraviolet photons to break materials（especially organic materials）or chemical bonds in the mediums around, to destroy materials in the non-hot process. 冷加工使用高能量的紫外线破坏材料（尤其是有机材料）或在介质中的化学键，从而破坏无热加工中的材料。	Cool working uses high road energy ultraviolet photons to break materials（especially organic materials）or chemical bonds in the mediums around, to destroy materials in the non-hot process.	文字	
	Cool working processing has special significance for laser marking, because it does not work through hot working, and will not produce hot damage or break cold stripping of chemical bonds, therefore, no hot deformation will be caused on the surface of the materials. 冷加工对于激光打标具有特别的意义，因为它不通过热加工，不会产生热损伤，也不会打破化学键的冷剥离。因此，材料表面也不会变形。			
2. Advantages 优点	Advantages 优点 Compared with other marking technologies such as ink jet printing and mechanical marking, laser marking has a number of advantages. 与其他的打标技术如喷墨打印和机械打标等相比，激光打标有很多优势。			
	Such as very high processing speeds, low operation cost（no use of consumables）, constant high quality and durability of the results, avoiding contaminations, the ability to write very small features, and very high flexibility in automation. 比如较高的加工速度、低操作成本（不需要使用耗材）、持续高质量以及耐久性能，同时避免了污染，能够标记很小的特征，以及在自动操作中有较高的灵活性。	Advantages： High processing speeds； Low operation cost（no use of consumables） Constant high quality； Durability of the results Avoiding contaminations Ability to write very small features High flexibility in automation		

标题	专业内容（讲稿与中英字幕）	PPT 中需显示的文字	表现形式	素材
	The physical principle of laser marking is based on light amplification by simulated emission of radiation. 激光打标的物理原理是基于受激辐射的光放大。	The physical principle of laser marking is based on light amplification by simulated emission of radiation.		图 2　激光打标原理图
	In other words, the energy emitted by a source（e. g. arc lamp）onto an active source（e. g. Nd：YAG crystal）will be collected and concentrated between two opposing mirrors. 换言之，由源极（如弧光灯）发出照到放射源（如掺钕的钇钕石榴石晶体）之上能量被集中在两面对立的镜子中。		对照原理图讲解	图 2　激光打标原理图
3. The physical principle of laser marking 激光打标的物理原理	**Part 1** The oscillation of specific light particles（photons）that move between the mirrors generates a high power laser beam. Only part of this beam passes through the output mirror（typically 10% – 20%），the rest being reinjected for the amplification process：the laser beam is thus created. 在镜子间移动的特殊光微粒（光子）的振动产生了高能量的激光束。这一光束，一部分穿过输出镜（通常是 10% ~20%），剩余的光束再次射入参与放大光束的过程，激光束因而产生了。	**Part 1** The oscillation of specific light particles（photons）that move between the mirrors generates a high power laser beam. Only part of this beam passes through the output mirror（typically 10% – 20%），the rest being reinjected for the amplification process：the laser beam is thus created.	强调图上的 Part 1	图 2　激光打标原理图
	Part 2 The beam is deflected by two scanning mirrors, driven electronically and via software, thus enabling a very high scanning speed and exceptional positioning accuracy. 两面振镜使激光束偏转，通过电驱动和软件，可以产生很高的扫描速度和惊人的定位准确性。	**Part 2** The beam is deflected by two scanning mirrors, driven electronically and via software, thus enabling a very high scanning speed and exceptional positioning accuracy.	强调图上的 Part 2	图 2　激光打标原理图

标题	专业内容（讲稿与中英字幕）	PPT 中需显示的文字	表现形式	素材
3. The physical principle of laser marking 激光打标的物理原理	Part 3 The flat field lens then concentrates the beam energy in a very tiny spot and greatly increases the power density on the surface to be marked. 场镜在很小的点上聚集光束能量，并在需要打标的材料表面增加能量密度。	Part 3 The flat field lens then concentrates the beam energy in a very tiny spot and greatly increases the power density on the surface to be marked.	强调图上的 Part 3	图 2　激光打标原理图
4. Three categories 三类	The different markings can be classified in three categories： Engraving with material removal （sublimation） Material annealing with surface color change Marking through layer removal. 打标可以分为三类： 通过材料切除来雕刻（升华） 通过退火使材料表面颜色发生改变 通过切除涂层进行打标。	The different markings can be classified in three categories： Engraving with material removal； （sublimation） Material annealing with surface color change； Marking through layer removal.	3 张图并列显示	图 3　通过材料切除来雕刻（升华） 图 4　通过退火使材料表面颜色发生改变 图 5　通过切除涂层进行打标
	Engraving with material removal （sublimation）： The energy is delivered with high peak power pulses so that the material is instantaneously removed without thermal side-effects on the parts. 通过材料切除来雕刻（升华）： 能量由高峰值能量脉冲发出，因此材料可以瞬间被移除，而不会在部件上产生热副效应。	Engraving with material removal （sublimation）		图 3　通过材料切除来雕刻（升华） Material surface ablation
	Material annealing with surface color change； The energy is delivered with lower pulses, heating the material and changing the surface appearance. No material is removed. 通过退火使材料表面颜色发生改变； 低脉冲产生能量，加热材料并改变表面外形，无须切除材料。	Material annealing with surface color change		图 4　通过退火使材料表面颜色发生改变 Surface color change
	Marking through layer removal： On coated material, the contrast is created by removing the top layer, thus showing the color of the base material. 通过切除涂层进行打标： 在有涂层的材料上，通过切除涂层顶层造成对比，因而显示出基底材料的颜色。	Marking through layer removal		图 5　通过切除涂层进行打标 One layer removed Two layer removed
	The end. Thank you！	The end. Thank you！		

Task V   Enhancing Competency

5.8   Everyday Use of Lasers

In everyday life, we're more or less surrounded by laser applications. Carpenters use laser instead of spirit levels, hunters use laser instead of ordinary telescopic sights and most likely, you use laser when you listen to music.

DVD

A DVD player (Fig. 5.16) contains a laser that is used not because it produces a parallel beam, but rather because the light emerges from a tiny point, which enables it to be focused on the different layers of the disk. By moving the lens sideways—laterally, it is possible to reach areas farther in or out on the disk. By moving the lens along the beam—longitudinally, different depths can be reached in the disk. The information, ones and zeros, is stored in several layers, and only one layer is to be read at a time. Every point on a particular layer is read during every revolution of the disk.

Fig. 5.16   DVD player

In order to make room for a lot of information on every disk, the beam has to be focused on an area as small as possible. This cannot be done with any other light source than a laser.

Speed Measurement Using Laser

The method the police use to measure car speed is based on a laser signal that is sent towards the target, as is shown in Fig. 5.17. This beam bounces back and is mixed with light that has not hit the car. The result is an oscillation—the same as when you tune a guitar—with higher frequency the faster the target moves.

The speed has to be measured straight from the front or from the back. If it is measured at an angle, the speed is underrated. This means that you cannot get false values that are too high.

The measurement is dependent on the car having something that reflects well. The license plate is perfect, as are different types of reflecting objects. Fogged surfaces are okay, but it will reduce the maximum distance.

Fig. 5.17   Speed measurement using laser

Laser Distance Meter

The primary users of laser distance meters today are surveyors and constructors, but the car industry is catching on. Least spectacular is the so-called parking assistance that helps the driver to estimate the distance to the car behind when parking. A more recent application measures the distance to the car in front of the driver when driving on highways or other roads. You simply lock in the distance to the car in front of you in order to maintain that distance. This makes driving more efficient and faster as long as it all works. This kind of laser is found in most robots with mechanical vision.

Optical Loudspeaker Cable

Any amplifier of worth nowadays has an optical cable for transmission to the loudspeakers. The advantage of this method is that it is insensitive to interference from electromagnetic fields, which is interference from electronic devices and radio transmitters such as cell phones. The light source used as a transmitter is a small laser semiconductor. All equipment using optic cable uses the same standard. For example, the maximum bit rate for broadband applications today is 50 – 100 times higher using optics, but the potential ratio is 10,000 times.

Exercise

Directions: Mark T (for True) if the statement agrees with the information given in the text; F (for False) if the statement contradicts the information given in the text.

(   ) 1. A DVD player contains a laser because the light emerges from a tiny point, which enables it to be focused on the different layers of the disk.

(   ) 2. In order to make room for a lot of information on every disk, the beam has to be focused on large space.

(   ) 3. When using laser measure speed, the speed should be measured at an angle, not from the front or from the back.

(   ) 4. Parking assistance helps the driver to estimate the distance to the car behind when parking.

(   ) 5. Optical loudspeaker cable is insensitive to interference from electromagnetic fields.

5.9　Laser Safety

Laser Beam Hazards

The laser produces an intense, highly directional beam of light. If directed, reflected, or focused upon an object, laser light will be partially absorbed, raising the temperature of the surface and the interior of the object, potentially causing an alteration or deformation of the material. Fig. 5. 18 shows the warning sign of laser radiation.

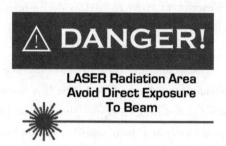

Fig. 5. 18　Warning sign

Lasers have the potential to damage both the eye and the skin. The human body is vulnerable to the output of certain lasers, and under certain circumstances, exposure can result in damage to the eye and skin. It is now widely accepted that the human eye is almost always more vulnerable to injury than human skin. The cornea, unlike the skin, does not have an external layer of dead cells to protect it from the environment. In the far-ultraviolet and far-infrared regions of the optical spectrum, the cornea absorbs the laser energy and may be damaged.

Non-beam Hazards

In addition to the direct hazards to the eye and skin from the laser beam itself, it is also important to address other hazards associated with the use of lasers.

These non-beam hazards, in some cases, can be life threatening, e. g. electrocution, fire, and asphyxiation. Because of the diversity of these hazards, It may be necessary for the employment of safety and industrial hygiene personnel to effect the hazard evaluations.

Safety Precautions Applicable to Solid-state Lasers

Enclosure of the beam and target in an opaque housing is the safest way of operating a laser (Fig. 5. 19). This level of safety precaution is mandatory for laser materials-processing systems operated in an industrial environment. In these systems interlocked doors, warning signs and lights, key-locked power switches, and emergency circuit breaker, and like precautions are taken to protect operators and passers-by from electrical and radiation hazards of the laser equipment. Also, microscopes and viewing ports are filtered or blocked to prevent the

Fig. 5. 19　Operating a laser

issuance of laser radiation, and laser impact points are surrounded by shields. At points of access for routine maintenance and set-up, warning signs are displayed prominently, and interlocks prevent firing of the laser while doors or ports are open.

In the laboratory, it is often not possible to enclose fully a high-power laser. In these situations, a series of safety precautions should be observed, e. g. wear laser safety glasses at all times! Never look directly into the laser light source or at scattered laser light from any reflective surface, etc.

Exercises

Directions: Choose the best answers from the four choices.

1. According to the first paragraph, laser light can _____ the surface of the object and alter the material.

A. evaporate

B. cool off

C. break

D. heat up

2. Which of the following statement is not true?

A. Lasers do damage to the eye rather than the skin.

B. The cornea has an external layer of dead cells to protect it from the environment.

C. The human body is easy to be destroyed by the output of lasers.

D. The cornea absorbs the laser energy in the far-ultraviolet and far-infrared regions of the optical spectrum and may be damaged.

3. What is the meaning of the word "mandatory" in the second line of Para 6?

A. Beneficial.

B. Effective.

C. Compulsory.

D. Imaginable.

4. To prevent the issuance of laser radiation, safety personnel should _____ .

A. filter and block the microscopes and viewing ports

B. show warning signs and light

C. protect operators from radiation hazards

D. prevent firing of the laser

5. In an industrial environment, the safest way of operating a laser is to _____ .

A. wear laser safety glasses at all times

B. never look directly into the laser light source

C. enclose the beam and target in an non-transparent housing

D. take precautions to protect passer-bys from electrical hazards

Unit 6　Fiber-optic Communication

In this unit, you are going to learn the following contents:

Task Ⅰ　Technical Principles and Device Cognition

6. 1　Brief Introduction of Fiber-optic Communication

6. 2　Typical Optical Transmitter—LEDs

6. 3　Typical Optical Receivers

Task Ⅱ　Technology Application Understanding

6. 4　Comparison with Electrical Transmission

Task Ⅲ　Skill Honing

6. 5　How to Translate with Aid of the Internet and Software

Task Ⅳ　Job Task Challenges

6. 6　Career Speaking: Business Meeting

6. 7　Writing: Memo

Task Ⅴ　Output and Evaluation

Task Ⅵ　Knowledge Expansion

6. 8　Optic Fibers

6. 9　Fiber to the Home

Unit 6

Fiber-optic Communication

Task I   Technical Principles and Device Cognition

6. 1   Brief Introduction of Fiber-optic Communication

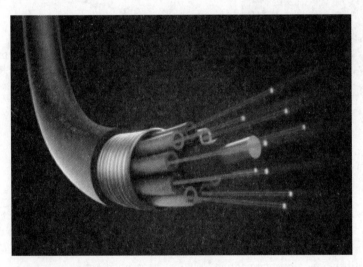

Learning Objectives

In this section, you will:

1. understand the meaning of fiber-optic communication.
2. learn about the development of optical fiber communication.
3. comprehend how the fiber-optic communication works.
4. learn the advantages and obstacles of optical fiber compared with copper wire.

Warm-up Activity

Directions：Fig. 6. 1 shows us common communication devices in our daily life. Please write down what they transmit? You may work with your partner.

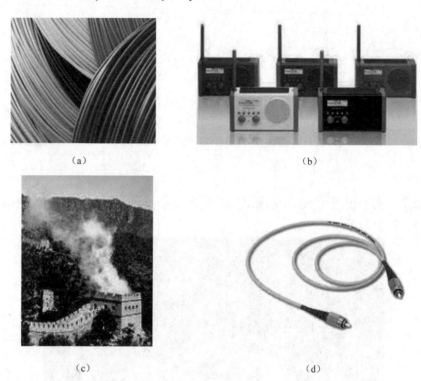

Fig. 6. 1　Common communication devices
（a）Cables；（b）Radio；（c）Beacon tower；（d）Fibers

Text

What is fiber-optic communication? It is a method of transmitting information from one place to another by sending pulses of light through an optical **fiber**.

Optical fiber was successfully developed in 1970 by Corning Glass Works, with attenuation low enough for communication purposes (about 20 dB/km), and at the same time, GaAs semiconductor lasers were developed that were compact and therefore suitable for transmitting light through fiber optic cables for long distances.

After a period of research starting from 1975, the first commercial fiber-optic communication system was developed, which operated at a wavelength around 0. 8 μm and used GaAs semiconductor lasers.

fiber *n.* 纤维；光纤

This first-generation system operated at a bit rate of 45 Mb/s with **repeater** spacing of up to 10 km.

The light forms an electromagnetic carrier wave that is **modulated** to carry information. First developed in the 1970s, fiber-optic communication systems （Fig. 6.2） have revolutionized the telecommunications industry and have played a major role in the **advent** of the information age. Because of its advantages over electrical transmission, optical fibers have largely replaced copper wire communications in core networks in the developed world.

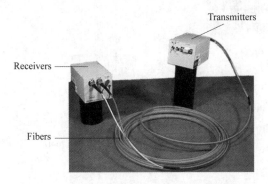

Fig. 6.2　Fiber-optic communication

And how exactly does the fiber-optic communication works? Simply speaking, there are 3 steps in short.

（1） Use an **optical transmitter** to create the optical signal.

（2） Relay the signal along the fiber, make sure the signal does not become too **distorted** or weak.

（3） Use a **receiver** to receive the optical signal, and **converting** it into an electrical signal.

Optical fiber is used by many telecommunications companies to transmit telephone signals, Internet communication, and **cable television** signals. Because much lower attenuation and interference, optical fiber has large advantages over existing copper wire in long-distance and high-demand applications. However, **infrastructure** development within cities was relatively difficult and time-consuming, and fiber-optic systems were complex and expensive to install and operate. Due to these difficulties, fiber-optic communication systems have primarily been installed in long-distance applications, where they can be used to their full transmission capacity, **offsetting** the increased cost. Since 2000, the prices for fiber-optic communications have dropped considerably. The price for rolling out

repeater *n.* 中继器；转发器

modulate *vt.* （信号）调制

advent *n.* 出现或到来

optical transmitter ［通信］光发送机；光透射器

distorted *adj.* 歪曲的；受到曲解的

receiver *n.* 接收器

convert *vt.* 使转变；转换……

cable television 电缆电视，有线电视

infrastructure *n.* 基础设施；公共建设

offset *vt.* 抵消；弥补

fiber （Fig. 6. 3） to the home has currently become more **cost-effective** than that of rolling out a copper based network.

cost-effective *adj.* 划算的；成本效益好的

Fig. 6. 3　Rolling out fiber

Since 1990, when optical-amplification systems became commercially available, the telecommunications industry has laid a vast network of intercity and transoceanic fiber communication lines. By 2002, an **intercontinental** network of 250, 000 km of submarine communications cable with a capacity of 2. 56 Tb/s was completed, and although specific network capacities are **privileged** information, telecommunications investment reports indicate that network capacity has increased dramatically since 2004.

intercontinental *adj.* 洲际的；大陆间的

privileged *adj.* 特许的；专用的

 Get to Work

Ⅰ. Matching

Directions：*Match the words or expressions in the left column with the Chinese equivalents in the right column.*

1. optical fiber
2. optical-amplification
3. interference
4. attenuation
5. modulate
6. distort
7. cable
8. transmitter
9. receiver
10. electromagnetic

A. 光纤
B. 调制
C. 电缆
D. 电磁的
E. 发射器
F. 使失真
G. 接收器
H. 衰减
I. 干扰
J. 光放大器

II. Understanding Checking

Directions：Mark Y (for YES) if the statement agrees with the information given in the text；N (for NO) if the statement contradicts the information given in the text.

(　　) 1. Fiber-optic communication system is barely used in Internet communication.

(　　) 2. Fiber-optic communication system includes transmitter and receiver.

(　　) 3. Too expensive to install and operate is the reason that fiber-optic communication systems been installed in long-distance applications.

(　　) 4. Fiber-optic communication system was first developed in the 1970s.

III. Passage Completion

Directions：Fill in each of the blanks with one of the words or expressions in the box, making changes if necessary.

formed	operating	became	took over
went into	led to	merged	in

Advanced Fibreoptic Engineering (AFE) was 1. _____ in June 2005.

However, a fibreoptic business has been 2. _____ in our Witney premises since 1994. Some customers may even remember the same business running from "The Ropery" in Burford before that.

The initial business in Burford traded as GCA Electronics and distributed electronic components. One line of products, optical fibre coupling LEDs for ABB Hafo 3. _____ Sweden, 4. _____ a demand for fibre alignment. This manufacturing element soon dominated the distribution business and distribution was discontinued. The company 5. _____ GCA Fibreoptics.

In 1998, Uniphase 6. _____ with JDS Fitel to form JDS Uniphase (JDSU).

During 2001, a management buy out (MBO) was formed and Afonics Fibreoptics 7. _____ the Witney facilities, Yarnton being far too large. Afonics grew from 2001 through to 2005, but never achieved profitability. In May 2005, Afonics Fibreoptics 8. _____ administration.

IV. Translation

Directions：Complete the sentences by translating into English according to the Chinese given in brackets.

1. Relay the signal along the fiber, _____ (确保信号不会失真或衰减).

2. _____ (因为相较很低的失真的干扰), optical fiber has large advantages over existing copper wire in long-distance and high-demand applications.

3. Optical fiber is used by _____ (电信公司传输电话信号，互联网通信，和有线电视信号).

4. _____ (铺设光纤的价格目前已经越来越具有性价比) than that of rolling out a copper based network.

5. By 2002, an intercontinental network of 250, 000 km of _____ _____ (传输速度为 2.56 Tb/s 的海底通信电缆) was completed.

V. Sentence Remaking

Directions:*Simulate the following sentence patterns according to an example provided from the text.*

1. … have revolutionized … and have played a major role in …

E. g. Fiber-optic communication systems have revolutionized the telecommunications industry and have played a major role in the advent of the Information Age.

_____ have revolutionized _____ and have played a major role in _____.

2. … is used by … to …

E. g. Optical fiber is used by many telecommunications companies to transmit telephone signals.

_____ is used by _____ to _____.

3. Due to …, … has large advantages over …

E. g. Due to much lower attenuation and interference, optical fiber has large advantages over existing copper wire in long-distance and high-demand applications.

Due to _____, _____ has large advantages over _____.

4. … been installed in … where …

E. g. Fiber-optic communication systems have primarily been installed in long-distance applications, where they can be used to their full transmission capacity, offsetting the increased cost.

_____ been installed in _____ where _____.

5. Because …, … has large advantages over …

E. g. Because much lower attenuation and interference, optical fiber has large advantages over existing copper wire in long-distance and high-demand applications.

Because _____, _____ has large advantages over _____.

6.2　Typical Optical Transmitter—LEDs

Learning Objectives

In this section, you will:

1. learn about the difference between LEDs and laser diodes.

2. comprehend the operational principle of LEDs.

3. master the advantages of LEDs.

4. find the applications of LEDs.

Warm-up Activity

Directions: Can you identify the positive pole and the negative pole of the device in Fig. 6.4 (a)? Do you know how to install it at the circuit in Fig. 6.4 (b) to make the device give out light? Please draw it on your book.

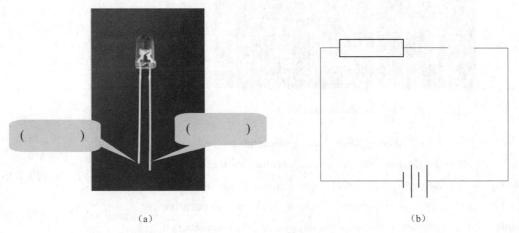

(a)　　　　　　　　　　　(b)

Fig. 6.4　Positive and negative

Text

The most commonly-used optical transmitters are semiconductor devices such as light emitting diodes （Fig. 6.5） and laser diodes （LD）. The difference between LEDs and laser diodes is that LEDs produce **incoherent light**, while laser diodes produce **coherent light**. For use in optical communications, semiconductor optical transmitters must be designed to be compact, efficient, and reliable, while operating in an optimal wavelength range, and directly modulated at high **frequencies**.

In its simplest form, an LED is a forward-biased **PN junction**, emitting light through **spontaneous emission**, a **phenomenon** referred to as **electroluminescence**. The emitted light is incoherent with a relatively wide spectral width of 30 – 60 nm. Communications LEDs are most commonly made from GaAsP or GaAs. Because GaAsP LEDs operate at a longer wavelength than GaAs LEDs （1.3 μm vs. 0.81 – 0.87 μm）. The large spectrum width of LEDs causes higher fiber dispersion, considerably limiting their bit rate-distance product （a common measure of usefulness）. LEDs are suitable primarily for local-area-network applications with bit rates of 10 – 100 Mb/s and transmission distances of a few kilometers.

incoherent light 非相干光
coherent light 相干光

frequency *n.* 频率；频繁
PN junction PN 结
spontaneous emission 自发发射
phenomenon *n.* 现象
electroluminescence *n.* ［电子］场致发光，电致发光

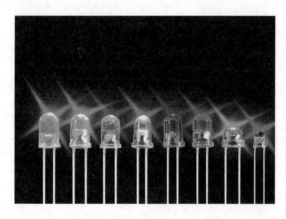

Fig. 6.5　Light emitting diodes

The low energy **consumption**, low **maintenance** and small size of modern LEDs has led to uses as **status indicators** and displays on a variety of equipment and installations. Large-area LED displays are used as **stadium** displays and as **decorative** displays. Thin, lightweight message displays are used at airports and

consumption *n.* 消费；消耗
maintenance *n.* 维护，维修
status indicators 状态指示器
stadium *n.* 体育场；露天大型运动场
decorative *adj.* 装饰性的；装潢用的

railway stations, and as destination displays for trains, buses, **trams**, and **ferries**.

With the development of high efficiency and high power LEDs, it has become possible to use LEDs in lighting and **illumination**. Replacement light **bulbs** have been made. LEDs are used as street lights and in other **architectural** lighting (Fig. 6. 6) where color changing is used. The **mechanical robustness** and long lifetime is used in automotive lighting on cars, motorcycles and on bicycle lights. LED street lights are used on poles and in parking garages. In 2007, the Italian village Torraca was the first place to convert its entire illumination system to LEDs.

tram *n.* 电车轨道
ferry *n.* 渡轮；摆渡

illumination *n.* 照明
bulb *n.* 电灯泡
architectural *adj.* 建筑学的；建筑上的
mechanical *adj.* 机械的；力学的
robustness *n.* 稳健性；健壮性

Fig. 6. 6　LEDs are used in architectural lighting

LEDs are used in **aviation** lighting. Airbus has used LED lighting in their Airbus A320 Enhanced since 2007, and Boeing plans its use in the 787. LEDs are also being used now in airport and **heliport** lighting. LED airport **fixtures** currently include medium-intensity runway lights, runway centerline lights, **taxiway** centerline & edge lights, guidance signs and obstruction lighting. LEDs are also suitable for backlighting for LCD televisions and lightweight **laptop** displays and light source for DLP projectors (See LED TV). RGB LEDs raise the color **gamut** by as much as 45%. Screens for TV and computer displays can be made thinner using LEDs for backlighting.

aviation *n.* 航空

heliport *n.* 直升机机场
fixture *n.* 设备；固定装置
taxiway *n.* 滑行道

laptop *n.* 膝上型轻便计算机
gamut *n.* 整个范围

Get to Work

I. Matching

Directions：*Match the words or expressions in the left column with the Chinese equivalents in the right column.*

1. semiconductor
2. light emitting diodes
3. laser diodes
4. incoherent lights
5. frequencies
6. electroluminescence
7. illumination
8. laptop
9. energy consumption
10. phenomenon

A. 现象
B. 能量损耗
C. 膝上型轻便计算机
D. 照明
E. 场致发光
F. 非相干光
G. 半导体二极管
H. 半导体
I. 频率
J. 发光二极管

II. Understanding Checking

Directions：*Mark Y (for YES) if the statement agrees with the information given in the text; N (for NO) if the statement contradicts the information given in the text.*

(　　) 1. According to the text, LED is a kind of metal device.

(　　) 2. GaAsP LEDs operates at a wavelength of $0.81 - 0.87$ μm.

(　　) 3. LEDs are suitable for short distance transmission.

(　　) 4. Due to LED has different kind of colors, they are used in lighting and illumination.

III. Passage Completion

Directions：*Fill in each of the blanks with one of the words or expressions in the box, making changes if necessary.*

over	comprised	bit	instead	from
over	since	weighed	similar	onto

LEDs have come a long way 1. _____ the early days of lighting up digital clock faces. In the 2000s, LCD TVs took 2. _____ the high definition market and represented a huge step over old standard definition CRT televisions. LCD displays were even a major step 3. _____ HD rear-projection sets that 4. _____ well over 100 pounds (45.4 kg). Now LEDs are poised to make a 5. _____ jump. While LCDs are far thinner and lighter than massive rear-projection sets, they still use cold cathode fluorescent tubes to over a white light 6. _____ the pixels that make up the screen. Those add weight and thickness to the television set. LEDs solve both problems.

Have you ever seen a gigantic flatscreen TV barely an inch thick? If you have, you've seen an LED television. Here's where the acronyms get a 7. _____ confusing：those LED TVs are still LCD TVs, because the screens themselves are 8. _____ of liquid crystals. Technically, they're

LED-backlit LCD TVs. 9. _____ of fluorescent tubes，LEDs shine light 10. _____ behind the screen，illuminating the pixels to create an image.

Ⅳ. Translation

Directions：Complete the sentences by translating into English according to the Chinese given in brackets.

1. _____ (半导体光发射器，例如发光二极管和激光二极管) is the most common optical transmitter in our daily life.

2. LED is wildly used in TV because _____ (红绿蓝颜色表示法将颜色范围提高了45%)。

3. The light given off by LED _____ (是光谱范围为 30 ~ 60 nm 的不相干光).

4. With the development of high efficiency and high power LEDs is _____ (有可能应用在照明上)。

5. Maybe someday in the further，we could _____ (将家庭照明系统换成 LED)。

Ⅴ. Sentence Remaking

Directions：Simulate the following sentence patterns according to an example provided from the text.

1. … in the advent of …

E. g. Fiber-optical communication systems have played a major role in the advent of the information age.

_____ in the advent of _____.

2. The difference between … is that …

E. g. The difference between LEDs and laser diodes is that LEDs produce incoherent light, while laser diodes produce coherent light.

The difference between _____ is that _____.

3. … has led to …

E. g. The low energy consumption，low maintenance and small size of modern LEDs has led to uses as status indicators and displays.

_____ has led to _____.

4. With the development of … it has become possible to …

E. g. With the development of high efficiency and high power LEDs it has become possible to use LEDs in lighting and illumination.

With the development of _____ it has become possible to _____.

5. … was the first to …

E. g. The Italian village Torraca was the first place to convert its entire illumination system to LEDs.

_____ was the first to _____ .

6.3   Typical Optical Receivers

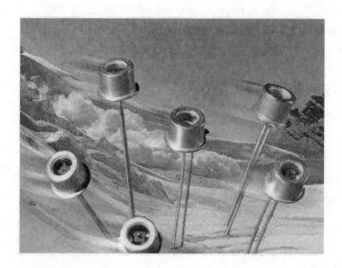

Learning Objectives

In this section, you will:

1. learn about the main component of optical receiver.

2. compare several types of photodiodes and find their applications.

Warm-up Activity

Directions: Which of the devices in Fig. 6.7 contain optical receivers? Discuss with your partner and orally explain your reasons.

(a)   (b)

Fig. 6.7   Devices contain optical receivers

(a) TV set; (b) IC card analyzer

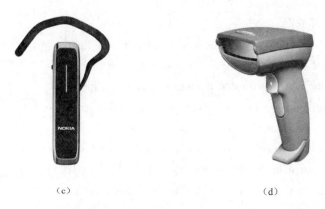

（c）   （d）

Fig. 6. 7   Devices contain optical receivers（Continued）

（c）Bluetooth headset；（d）Barcode scanner

 Text

As we all know, light can be detected by eyes. But eye is not suitable for modern fiber communications because its response is too slow, its sensitivity to low-level signals is **inadequate**, and it is not easily connected to electronic receivers for amplification, **decoding**, or other **signal processing**. Furthermore, the spectral response of the eye is limited to wavelength between 0. 4 μm and 0. 7 μm, where fibers have high loss.

inadequate *adj.* 不充分的，不适当的

decoding *n.* ［通信］解码

signal processing 信号处理

The main component of an optical receiver is a photodetector, which converts light into electricity using the photoelectric effect. The photodetector is typically a semiconductor-based **photodiode** （Fig. 6. 8）. Several types of photodiodes include PN photodiodes, A PIN photodiodes, and **avalanche photodiodes**.

photodiode *n.* ［电子］光电二极管

avalanche photodiode 雪崩光电二极管

Fig. 6. 8   photodiodes

PIN diodes are much faster and more sensitive than ordinary PN junction diodes, and **hence** are often used for optical communications

hence *adv.* 因此；今后

and in lighting regulation.

PN photodiodes are not used to measure extremely low light intensities. Instead, if high sensitivity is needed, avalanche photodiodes, intensified **charge-coupled devices** or **photomultiplier tubes** are used for applications such as **astronomy**, **spectroscopy**, night vision equipment and laser range finding.

charge-coupled device 电荷耦合器

photomultiplier tube 光电倍增管

astronomy *n.* 天文学

spectroscopy *n.* ［光］光谱学

The PIN device is cheaper, is less sensitive to temperature, and requires lower **reverse bias** voltage than the APD. The speeds of the two devices are comparable, so the PIN diode is preferable in most systems. The APD gains its need when the system is less limited, as occurs for long-distance links. Metal-semiconductor-metal (MSM) photo detectors are also used due to their suitability for **circuit integration** in wavelength-division multiplexers.

reverse bias 反相偏压

circuit *n.* 电路，回路

integration *n.* 集成；综合

Photodiodes are used in consumer electronics devices such as compact disk players, smoke detectors, and the receivers for **remote controls** in VCRs and televisions.

remote control 遥控器

In other consumer items such as camera light meters, clock radios (the ones that **dim** the display when it's dark) and street lights, **photoconductors** (Fig. 6.9) are often used rather than photodiodes, although in principle either could be used.

dim *vt.* 使暗淡，使失去光泽

photoconductor *n.* ［物］光电导体；光电导元件；光敏电阻

Photodiodes are often used for accurate measurement of light intensity in science and industry. They generally have a better, more linear response than photoconductors.

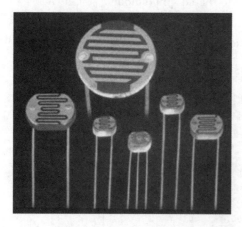

Fig. 6.9 Photoconductors

tomography *n.* ［医］X 线断层摄影术

computed tomography 计算机断层扫描

They are also widely used in various medical applications, such as detectors for **computed tomography** or instruments to analyze samples (**immunoassay**).

immunoassay *n.* 免疫分析，免疫测定

Get to Work

I. Understanding Checking

Directions: *Mark Y (for YES) if the statement agrees with the information given in the text*; *N (for NO) if the statement contradicts the information given in the text.*

(　　) 1. The eye is not suitable for modern fiber communication because it is not sensitive enough.

(　　) 2. The photodetector converts light into electricity.

(　　) 3. All the photodetectors are semiconductor-based photodiodes.

(　　) 4. The PIN device is preferable in most system because its speed is much faster than APD.

II. Matching

Directions: *Match the words or expressions in the left column with the Chinese equivalents in the right column.*

1. decoding	A. 光电二极管
2. photodetector	B. 集成
3. photoelectric effect	C. 回路
4. photodiode	D. 家用电器
5. avalanche photodiodes	E. 光谱响应
6. photoconductors	F. 光电导体
7. integration	G. 光电效应
8. consumer electronics	H. 雪崩光电二极管
9. spectral response	I. 解码
10. circuit	J. 光电探测器

III. Passage Completion

Directions: *Fill in each of the blanks with one of the words or expressions in the box, making changes if necessary.*

offering	tested	in	on	advice
instructions	access	savings	range	over

QPhotonics is a specialized distributor of semiconductor light emitting devices 1. _____ 660 – 1,800 nm wavelength 2. _____ for your research, development, and production. Each laser device is fully 3. _____ in house and characterized for performance. We focus 4. _____ establishing long-term relations with our customers by 5. _____ value in our products and service.

We offer efficient and inexpensive solutions to engineers and researchers looking for semiconductor light sources:

(1) Large variety (6. _____ 100 models);

(2) Rapid 7. _____ to the right device, without compromising specifications or quality;

(3) Detailed datasheets online;

(4) Fast delivery (most laser diodes are in stock);

(5) Price 8. _____ on standard products;

(6) Unusual wavelengths, custom configurations fabricated in small lots;

(7) 9. _____ on laser diode selection and operation;

(8) Complementary products (laser controllers, mounts, photodiodes);

(9) Simple operating 10. _____ and handouts;

(10) Efficient shopping experience.

We strive for customer satisfaction by providing fully tested quality products, comprehensive datasheets, and operating manuals.

IV. Translation

Directions: Complete the sentences by translating into English according to the Chinese given in brackets.

1. Consumer electronics devices such as compact disk players, smoke detectors _____ _____ (都使用了光电二极管).

2. The eyes' response is _____ (局限在 0.4 ~ 0.7 μm 的波段内).

3. PIN diodes often used for optical communications and in lighting regulation _____ _____ (因为它比普通的 PN 结器件速度更快且更灵敏).

4. The PIN device is cheaper, is less sensitive to temperature _____ _____ (并且与雪崩二极管相比只需较低的反向电压).

5. (由于光电二极管较光敏电阻有更好的线性和更好的响应) _____ _____, photodiodes are often used for accurate measurement of light intensity in science and industry.

V. Sentence Remaking

Directions: simulate the following sentence patterns according to an example provided from the text.

1. … is not suitable for … because …

E. g. The eye is not suitable for modern fiber communications because its response is too slow.

_____ is not suitable for _____ because _____ .

2. … than …, and hence…

E. g. PIN diodes are much faster and more sensitive than ordinary PN junction diodes, and hence are often used for optical communications and in lighting regulation.

_____ than _____, and hence _____ .

3. … are comparable, … is preferable in …

E. g. The speeds of the two devices are comparable, so the PIN diode is preferable in most systems.

_____ are comparable, _____ is preferable in _____ .

4. ... are used in ... such as ...

E. g. Photodiodes are used in consumer electronics devices such as compact disk players, smoke detectors.

_____ are used in _____ such as _____ .

5. ... in any condition.

E. g. We can't mix up the two kinds of devices in any condition.

_____ in any condition.

Task Ⅱ Technology Application Understanding

6.4 Comparison with Electrical Transmission

Learning Objectives

In this section, you will:

1. learn about the main benefits of fiber.

2. comprehend why in short distance and relatively low bandwidth applications, electrical transmission is often preferred.

3. find the important features of fiber when fiber may be used even for short distance or low bandwidth applications.

4. learn the installation of optical fiber cables.

Warm-up Activity

Directions: *Fill in the brackets in Fig. 6.10（b）with the words or expressions given in Fig. 6.10（a）according to the wavelength of each electromagnetic wave.*

X ray	X 射线
infrared	红外线
microwave	微波
gamma ray	伽马射线
ultraviolet（UV）rays	紫外线
wireless	无线电
visible light	可见光

（a）

Fig. 6.10 Wavelength

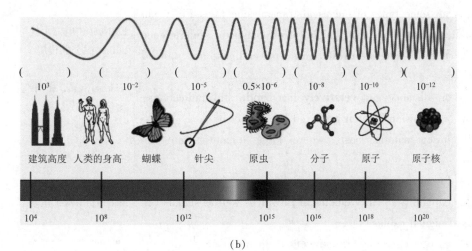

10^3 10^{-2} 10^{-5} $0.5×10^{-6}$ 10^{-8} 10^{-10} 10^{-12}

建筑高度   人类的身高   蝴蝶   针尖   原虫   分子   原子   原子核

10^4 10^8 10^{12} 10^{15} 10^{16} 10^{18} 10^{20}

（b）

Fig. 6. 10   Wavelength（Continued）

 Text

The choice between optical fiber（Fig. 6. 11）and electrical transmission for a particular system is made based on a number of **trade-offs**.

trade-off *n.* 权衡

Fig. 6. 11   Single-mode optical fiber in
an underground service pit

Optical fiber is generally chosen for systems requiring higher **bandwidth** or spanning longer distances than electrical cabling can accommodate.

bandwidth *n.*［电子］［物］带宽

The main benefits of fiber are its exceptionally low loss and its **inherently** high data-carrying capacity. Thousands of electrical links would be required to replace a single high bandwidth fiber cable. Another benefit of fibers is that even when run alongside each other for long distances, fiber cables experience effectively no **crosstalk**,

inherently *adv.* 内在地；固有地；天性地

crosstalk *n.* 串话干扰，串台

in contrast to some types of electrical transmission lines. Fiber can be installed in areas with high electromagnetic interference (**EMI**), such as alongside **utility** lines, power lines, and railroad tracks.

EMI *abbr.* 电磁干扰

utility *n.* 公共设施

In short distance and **relatively** low bandwidth applications, electrical transmission is often preferred because of its:

relatively *adv.* 相当地；相对地，比较地

(1) lower material cost, where large quantities are not required;

(2) lower cost of transmitters and receivers;

(3) capability to carry **electrical power** as well as signals (in specially-designed cables);

electrical power 电功率

(4) ease of operating **transducers** in linear mode.

transducer *n.* 传感器

In certain situations, fiber may be used even for short distance or low bandwidth applications, due to other important **features**:

feature *n.* 特色，特征

(1) **Immunity** to electromagnetic interference, including nuclear **electromagnetic pulses** (although fiber can be damaged by alpha and beta radiation).

immunity *n.* 免疫力；豁免权；免除

electromagnetic pulse 电磁脉冲

(2) High electrical **resistance**, making it safe to use near high-voltage equipment or between areas with different earth **potentials**.

resistance *n.* 阻力；电阻

potential *n.* [电]电势

(3) Lighter mass—important, for example, in aircraft.

(4) No sparks—important in flammable or explosive gas environments.

(5) Not electromagnetically radiating, and difficult to tap without disrupting the signal—important in high-security environments.

(6) Much smaller cable size—important where pathway is limited, such as networking an existing building.

Optical fiber cables can be installed in buildings with the same equipment that is used to install copper and **coaxial cables**, with some **modifications** due to the small size and limited pull tension and bend radius of optical cables. Optical cables can typically be installed in duct systems in **spans** of 6,000 m or more depending on the duct's condition, layout of the duct system, and installation technique. Longer cables can be **coiled** at an intermediate point and pulled farther into the duct system as necessary.

coaxial cable 同轴电缆

modification *n.* 修改，修正；改变

span *n.* 跨度；一段时间

coil *vt.* 盘绕，把……卷成圈

Get to Work

Ⅰ. Matching

Directions：*Match the words or expressions in the left column with the Chinese equivalents in the right column.*

1. trade-offs	A. 电功率
2. bandwidth	B. 免于，抗……性
3. crosstalk	C. 电阻
4. coaxial cables	D. 公共设施
5. potential	E. 特点
6. feature	F. 权衡
7. immunity to	G. 同轴电缆
8. electrical power	H. 带宽
9. resistance	I. 潜在的
10. utility	J. 串台

Ⅱ. Understanding Checking

Directions：*Mark Y（for YES）if the statement agrees with the information given in the text；N（for NO）if the statement contradicts the information given in the text.*

(   ) 1. Fiber-optic communication system's best benefit is its exceptionally low loss and its inherently high data-carrying capacity.

(   ) 2. Due to the fiber cables experience effectively no crosstalk, we can use it in the confidentiality communication.

(   ) 3. To lay a network system in an existing building, fiber cables is the better choice.

(   ) 4. Electrical transmission is often preferred in long distance transmission.

(   ) 5. Fiber-optic communication has taken the place of electrical transmission now.

Ⅲ. Passage Completion

Directions：*Fill in each of the blanks with one of the words or expressions in the box, making changes if necessary.*

deliver	refer	individual	during	looking
current	main	within	drive	increase

Stop and think how your Internet usage has evolved 1. _____ the last few years. If you're like most people, you're doing and expecting a lot more of your Internet like increased interactivity, rich media and uploading and downloading pictures and video.

More large files are moving across the cyberspace network these days, and experts expect that trend will only 2. _____ . A January 2008 study by the Discovery Institute estimates new technologies will 3. _____ Internet traffic up by 50 times its 4. _____ rate 5. _____ the next 10 years.

The pressure for better connectivity is one of the 6. _____ reasons providers and users are 7. _____ at fiber-to-the-home broadband connections as a potential solution.

Fiber-to-the-home broadband connections, or FTTH broadband connections, 8. _____ to fiber optic cable connections for 9. _____ residences. Such optics-based systems can 10. _____ a multitude of digital information—telephone, video, data, et cetera—more efficiently than traditional copper coaxial cable for about the same price. FTTH premises depend on both active and passive optical networks to function.

Ⅳ. Translation

Directions: Complete the sentences by translating into English according to the Chinese given in brackets.

1. _____ (消耗较少的发射器和接收器). is one of the reason that electrical transmission is often preferred in short distance applications.

2. _____ (一根光纤携带的信号量) equal to thousands of electrical links.

3. Fibers are preferred where pathway is limited _____ (因为它的尺寸较小).

4. If it is necessary _____ (我们能将光纤卷曲起来).

5. Lighter mass, _____ (使得光纤较铜线便于运输).

Ⅴ. Sentence Remaking

Directions: Simulate the following sentence patterns according to an example provided from the text.

1. The main benefits of ... Another benefit of ...

E. g. The main benefits of fiber are its exceptionally ... Another benefit of fibers is that even when run along...

The main benefits of _____. Another benefit of _____.

2. ... in spans of ...

E. g. Optical cables can typically be installed in duct systems in spans of 6,000 m or more.

_____ in spans of _____.

3. Ease of ...

E. g. Ease of operating transducers in linear mode.

Ease of _____.

4. ... has the capability to ...

E. g. Electrical transmission has the capability to carry electrical power as well as signals (in specially-designed cables)

_____ has the capability to _____.

5. ... in contrast to ...

E. g. Fiber cables experience effectively no crosstalk, in contrast to some types of electrical transmission lines. Fiber can be installed in areas with high.

_____ in contrast to _____.

Task Ⅲ　Skill Honing

6.5　How to Translate with Aid of the Internet and Software
如何借助网站及软件做英语翻译

随着 AI 技术的普及，人们可以借助网络和软件实现高质量且快速地翻译。以下文为例：

A technique that has had recent success is laser cooling. This involves atom trapping, a method where a number of atoms are confined in a specially shaped arrangement of electric and magnetic fields. Shining particular wavelengths of laser light at the ions or atoms slows them down, thus cooling them. As this process continued, they are all slowed and have the same energy level, forming an unusual arrangement of matter known as a Bose-Einstein condensate.

百度翻译的译文如下。

最近取得成功的一项技术是激光冷却。这涉及原子捕获，一种将许多原子限制在电场和磁场的特殊形状排列中的方法。向离子或原子照射特定波长的激光会减慢离子或原子的速度，从而使其冷却。随着这一过程的继续，它们都变慢了，具有相同的能级，形成了一种不寻常的物质排列，称为玻色－爱因斯坦凝聚态。

有道翻译的译文如下。

最近取得成功的一项技术是激光冷却。这涉及原子捕获，一种将许多原子限制在一个特殊形状的电场和磁场中排列的方法。用特定波长的激光照射离子或原子，使它们减速，从而使它们冷却。随着这一过程的继续，它们都变慢了，并具有相同的能量水平，形成了一种不寻常的物质排列，称为玻色－爱因斯坦凝聚体。

人工翻译如下。

激光冷却技术最近获得成功。该技术涉及原子捕集，即将众多原子限制在特殊排列的电磁场里。将特定波长的激光束投射到离子或原子上，将它们的速度放慢，从而冷却它们。随着这一进程的继续，它们的速度均放慢，并有相同的能级，形成了特别的物质组合，被称为玻色－爱因斯坦凝聚态。

综合比较三篇译文，可总结出借助 AI 技术的网络翻译具有如下特点。

（1）速度快：可以实现实时翻译，极大地提高了翻译的效率和速度。

（2）处理大量数据：机器翻译可以轻松处理大量的文本数据，在短时间内完成翻译，节约人力资源。

（3）准确性高：AI 可以通过强大的语义分析能力和机器学习算法来理解口语和书面表达中的含义，从而减少翻译的错误率。

尽管随着技术的不断发展，网络翻译的翻译质量逐渐提高，但在某些场景下仍然需要人工介入，尤其是专业领域的特定用法、文本上下文语境的正确推断以及情感和情绪的恰当表达等方面，无法取代人工翻译。

那么，如何正确应用网络及软件工具呢？

（1）充分认识网络及软件工具在翻译时的利弊，用其长而避其短。

（2）熟知不同翻译网站及翻译软件的特色，有区别地综合运用，如有道翻译（http：//fanyi. youdao. com）搜索功能强大，同一单词或词组的翻译，往往能提供多种选择答案。CNKI 翻译助手（http：//dict. cnki. net）基于中国知网数据库，其专业词汇的翻译具有一定的参考价值。

（3）翻译网站及翻译软件只能作为翻译的辅助工具，有时在翻译单词和短语时能提供一定参考价值的选择，但一定要结合上下文，先甄别后采用。

（4）译文的准确性、流畅性、专业性主要取决于译者的中英文功底、翻译经验和专业知识的积累。

以下是一些翻译网站及翻译软件。

google 翻译：http：//translate. google. cn

CNKI 翻译助手：http：//dict. cnki. net

有道翻译：http：//fanyi. youdao. com

词博词典：http：//www. cibo. cn

爱词霸：http：//www. iciba. com

金山词霸软件

金山快译软件

Challenge

Directions：Please translate the following passage from English into Chinese or vice versa by taking advantage of what has been learned. Translation websites and software may help you.

1. Laser printer.

Producing near-professional quality print, the laser printer operates like a copying machine—with one important difference that the copying machine produces its image by focusing its lens on paper, while the laser printer traces an image by using a laser beam controlled by the computer. Laser printers produce text and graphics with high resolution; however, the print quality is not as high as that obtained by a phototype machine used by commercial textbook publishers. There are many inexpensive laser printers that print in black and white on the market today, and it is a good choice because of its beautiful graphics capabilities.

2. 九头鸟激光集团简介。

1990 年成立之初，九头鸟激光集团只生产焊接激光机。经过 20 多年的发展，该集团生产的各种类型激光机工业用途已十分广泛，如激光打标、激光打孔、激光切割、激光热处理，并实现与欧美最新设计和技术同步升级。秉承"服务顾客、创造价值"的理念，集团的核心竞争力不断得到提升。截至目前，集团在国内外共售出 7 000 多套工业激光机。

Task Ⅳ   Job Task Challenges

6.6   Career Speaking：Business Meeting

Ⅰ. Warm Up

Model

（A：Chairman   B：Nancy   C：Ben   D：Ann）

A：Thank you for coming, everyone. It's really a pleasure to see you all here. Our main aim today is to discuss the time and mode of shipment. Would you like to make any comment on that?

B：Our customers hope we can ship the goods at once in order that they will catch up with the *season*.

C：We haven't received their **L/C** （letter of credit）. According to the last *negotiation*, we won't arrange the shipment until the L/C arrives. Nancy, you'd better make sure whether they've sent the L/C.

B：OK, I take your point. They didn't mention the mode of shipment. I think we should produce the possible solution first so as to negotiate with them next week.

C：Then, do we have the goods sent by sea, by air or by railway?

B：I would prefer to have goods carried by railway. It's simple and cheap.

A：That sounds reasonable. What do you think about this proposal, Ann?

D：Well, why don't we choose sea shipment? It's safer.

B：Sometimes there is no direct *vessel*, and it's easy to cause a delay.

C：That's exactly the way I feel. How about by air? Since our customers worry about the time of delivery, it's the most convenient to ship by air.

D：I don't agree unless they pay more *freight*.

A：It seems a complex matter. We'll stop here for today and leave that to another time. Can we set the date for the next meeting, please?

Notes

season	*n.* 旺季
L/C	信用证
negotiation	*n.* 谈判
vessel	*n.* 船
freight	*n.* 运费

Ⅱ. Matching

Directions: *Please choose the corresponding sentences from the box to fill in the blanks. Then,* *play roles with your partners.*

（A：Buddy　　B：Henry　　C：Christina　　D：Windy）

A：Good morning, everyone. If we are all here, let's get started. ＿＿＿＿＿＿＿＿＿＿.

B：In my opinion, it's necessary to redesign our office layout.

C：Right. ＿＿＿＿＿＿＿＿＿＿＿＿＿＿＿＿＿＿＿

D：I also think so. The office can be decorated with green plants.

A：＿＿＿＿＿＿＿＿＿＿＿＿＿＿＿＿＿＿＿＿＿

D：Frankly speaking, most of them are old-fashioned and out of order, and this would lower the efficiency of work. ＿＿＿＿＿＿＿＿＿＿＿＿＿

C：Unfortunately, I see it differently. It is a waste of money. You know, we have not enough funds.

D：＿＿＿＿＿＿＿＿＿＿＿＿＿＿＿＿＿＿＿＿＿

S1：What do you think of present office facilities?

S2：Why don't we update or replace them?

S3：Our main aim today is to study how to improve the working condition.

S4：But, at least the problem will be solved step by step.

S5：We need new environment for a change to get refreshed.

Ⅲ. Closed Conversation

Directions: *Please fill in the blanks according to the Chinese. Then,* *play roles with your* *partners.*

（A：Chairman　　B：Lisa　　C：Frank　　D：Paul　　E：Molly）

A：＿＿＿＿＿＿＿＿＿＿＿＿＿＿＿＿＿＿＿＿＿＿＿＿＿

＿＿＿＿＿＿＿＿＿＿＿＿＿＿＿＿＿＿＿＿＿＿＿＿＿

（首先我欢迎你们并且感谢你们的到来，尤其是收到通知你们就来了。我知道你们都很忙，能抽出时间来参加这次会议是很不容易的。）We're here today to discuss the matter of packing. Any suggestions on packing are greatly appreciated.

B：＿＿＿＿＿＿＿＿＿＿＿＿＿＿＿＿＿＿＿＿＿＿＿＿＿

（在我看来，漂亮的包装有助于推销产品，加深消费者对我们品牌的印象。）So our packing should be exquisite in design.

D：I agree. Consumers are easy to be attracted by beautiful appearance.

A：＿＿＿＿＿＿＿＿＿＿＿＿＿＿＿＿＿＿＿（你是怎么想的，弗兰克?）

C：I think the key point of packing lies in protecting the goods from moisture and shock.

D：＿＿＿＿＿＿＿＿＿＿＿＿＿＿＿＿＿＿（用木箱怎么样?）

B：The cost will be much higher. ＿＿＿＿＿＿＿＿＿＿＿＿＿＿（我更青睐纸箱。）They are cheaper, lighter and easier to handle.

C：That's quite true, but I'm afraid the cardboard boxes are not strong enough for transportation.

E：May I have a word?

A：Of course, Molly. _____（请继续。）

E：I suggest foam plastics and an inner waterproof lining should be applied to protect the goods. In addition, we can reinforce the packing with wooden straps.

A：_____

（谢谢大家的好主意，我们在下周的会议上再继续讨论。）

IV. Semi-open Conversation

Directions：Please fill in the blanks with your own words. Then, play roles with your partners.

（A：Chairman　　B：Mary　　C：Sam　　D：May）

A：Let's get this meeting started. Is everyone here?

B：I don't think Peter is here.

A：Okay. We can't wait. _____

First, please explain briefly what your departments accomplished during the three weeks. Sam, have you finished the financial report yet?

C：Yes. _____

A：Good. How about marketing department, Mary?

B：We have been developing our new marketing campaign.

A：_____

B：Yes. We've opened up a new outlet.

A：Well done! May, you are the head of Customer Service Department. _____

D：My department has been doing a survey and collecting all the necessary data, but some customers refuse to cooperate.

A：Why?

D：I'm not sure. Sue is looking into that.

A：_____

E：Yes, I will.

A：Good luck to you. _____

V. Live Show

Challenge 1

Directions：Lucy holds a business meeting. Betty, Susan, Roy and Martin attend it. They discuss the subject of packing and shipping of products. Please act it out with your partner or partners. The model and the useful expressions below are just for your reference.

Challenge 2

Directions：Please hold a meeting to discuss the arrangement of receiving American guests. Act out in groups. The model and the useful expressions below are just for your reference.

Model

(A：Ann　　B：Bob　　C：Clair　　D：David　　E：Emily)

A：Good morning, everyone. Please be seated and we'll get started. As you can see here on the *agenda* we will be mainly talking about three issues：One, the designs of a new advertising *brochure*；Two, the approaches to cost reduction；and, Three, the improvement of the customer service. Now, let's begin with the first issue. Emily, what's your opinion of the designs of our present advertising brochure?

E：I'm fed up with the old brochure. The information's out of date, and the telephone numbers are all wrong.

A：Right. So would it help to think what we want to put in the brochure?

C：It should give a brief introduction to the new product and indicate its quality and features. A few photographs and an approach to contact are added to it. Simple as that.

D：I was thinking that the brochure should be a bit more "*visual*". We can use lots of photos of the product in use, the manufacturing process, satisfied customers and something else. More data and fewer words will have a better effect. What do you think?

E：I'm with you. In order to guide our *campaigns* to success in the new market, the designs of the brochure must be *novel* and *appealing* to eyes.

A：Could you put it in detail, Emily?

E：I am not an expert on designing. In my view, we can contact a design agency and get it to use our rough ideas to come up with a few *alternatives*.

B：That will cost a lot of money, I suppose.

E：But for a new company, such expenses are absolutely necessary.

D：I follow you, Emily. It's worth doing as long as the result is satisfactory.

A：I never thought of that. Maybe it works. Clair, what do you think about this proposal?

C：That's a good idea. We can have a try.

A：Well, David, would you like to contact several design agencies and get *quotations*?

D：Yes, no problem.

A：Bob and Clair, you two are in charge of the text for the brochure. OK?

B and C：OK.

Notes

agenda	*n.*	议事日程
brochure	*n.*	宣传手册
visual	*adj.*	直观的，视觉的
campaign	*n.*	广告计划
novel	*adj.*	新颖的
appealing	*adj.*	吸引人的
alternative	*n.*	选择
quotation	*n.*	估价

Useful Expressions

（1）开始会议。

Let's begin.

Let's get down to the business.

Shall we begin?

（2）表达意见。

I (really) feel that …

In my opinion …

If you ask me, I tend to think that …

（3）表示同意。

I totally agree with you.

That's (exactly) the way I feel.

I couldn't agree more.

（4）表示异议。

Up to a point I agree with you, but …

I'm sorry, but I'm not sure I agree.

(I'm afraid) I can't agree.

（5）要求发言。

May I have a word?

Can I say something here?

Excuse me for interrupting.

（6）询问意见。

Can we get your view?

How do you feel about …?

Has anyone else got anything to contribute?

（7）结束会议。

We'll stop here for today.

Let's call it a day.

If there are no other issues to discuss,

I'd like to wrap up this meeting.

6.7　Writing：Memo

备忘录（memo，是 memorandum 的简称），是以书面形式来传递内部各种事务信息的一种方式，可以起到通知变化、提醒事项、记录电话或会议内容等作用。备忘录结构明了，语言简洁，方便阅读与传递。其格式一般较为固定，通常分为开头和正文两个部分。开头包括 To：（收信人姓名、职务）、From：（发信人或留言人姓名、职务）、Date：（日期）和 Subject：（主题）。其格式如下：

Memo

To：

From：

Date：

Subject：

在写备忘录时要注意：

表达清晰，结构紧凑；

开门见山，无须寒暄；

一般不采用信头称呼的方式。

Sample 1

Memo

To：John Henry, Sales Manager

From：Sue Clinton, Saleswoman from ABC Foreign Trade Company

Date：June 8, 2011

Subject：Shipment of the Scanners

I'm calling about the shipment of the scanners coming from Nanjing. Our customers are badly in need of the goods. Please call me back at 695 – 8942 before this Friday.

Sample 2

Memo

To：All Staff

From：Jane Jackson, Marketing Manager

Date：May 30, 2011

Subject：Meeting on Marketing and Promotion

We'll have a meeting in the conference room at 10：00 a. m. on June 5. The purpose of the meeting is to discuss how to increase market shares and how to promote new products. Any suggestions and ideas are welcome.

Useful Expressions

I have several proposals for …

We'll discuss … at the meeting.

Please let me know your reactions to the plans.

Here we announce the appointment of …

I have the pleasure of announcing that …

Challenge

Directions: *On the morning of July 2, 2023, Wang Lili, a clerk from Wuhan Import & Export Company, calls Fang Xin, marketing manager of your company, but he is not available. Wang Lili would like to make an appointment with Fang Xin to talk about the details of shipment. She will call him again this Thursday. Please write a memo.*

Task V　Output and Evaluation

Fiber-optic Communication					
Target	Understand and describe Fiber-optic Communication system				
Requirement	Make an English presentation according to Unit 6. Draw a mind map. Produce Micro-video to express professional knowledge. Add accurate Chinese and English subtitles to the video				
Contents	Fiber-optical Communication				
Group: _____	**Project**	**Name**	**Software**	**Score**	**Requirements**
	Mind map				The logic should be sound, and the keywords used should be accurate.
	Script				The format should adhere to the specified guidelines, and it should cover all necessary contents.
	PPT				It should be consistent with the logic presented in the mind map. Clear and concise content/Consistent formatting/Limited text/Engaging graphics and animations.
	Speaking				Clear/fluent
	Subtitle				They should be bilingual, correct and synchronized with the student's speech.
	MP4				The video should effectively convey knowledge, be accurate, and visually engaging.
Operating environment	Win7/Win8/Win9/Win10/Win11/Mobile phone				
Product features					

Script Reference（略）

Task Ⅵ   Knowledge Expansion

6.8   Optical Fibers

An optical fiber (Fig. 6.12) consists of a core, **cladding**, and a **buffer** (a protective outer coating), in which the cladding guides the light along the core by using the method of total internal reflection. The core and the cladding (which has a lower-refractive-index) are usually made of high-quality **silica** glass, although they can both be made of plastic as well. Connecting two optical fibers is done by fusion splicing or mechanical splicing and requires special skills and interconnection technology due to the microscopic precision required to align the fiber cores.

Fig. 6.12   Optical fiber

Two main types of optical fiber used in optic communications include multi-mode optical fibers and single-mode optical fibers. A multi-mode optical fiber has a larger core ($\geqslant 50$ μm), allowing less precise, cheaper transmitters and receivers to connect to it as well as cheaper connectors. However, a multi-mode fiber introduces multimode distortion, which often limits the bandwidth and length of the link. Furthermore, because of its higher **dopant** content, multi-mode fibers are usually expensive and exhibit higher attenuation. The core of a single-mode fiber is smaller (< 10 μm) and requires more expensive components and interconnection methods, but allows much longer, higher-performance links. In order to package fiber into a commercially-viable product, it is typically protectively-coated by using ultraviolet, light-cured **acrylate polymers**, then terminated with optical fiber connectors, and finally assembled into a cable.

After that, it can be laid in the ground and then run through the walls of a building and deployed aerially in a manner similar to copper cables. These fibers require less maintenance than common twisted pair wires, once they are deployed.

Notes

cladding	*n.* 包层
silica	*n.* 硅石
dopant	*n.* 掺杂物
buffer	*n.* 保护层
fusion	*n.* 融化
acrylate polymers	*n.* 丙烯酸酯聚合物

Exercise

Directions: *Match the words or expressions about the advantage of the fibers in the left column with the expressions about the application of the fibers in the right column.*

1. Less signal degradation　　　　　　A. useful in computer networks

2. Non-flammable　　　　　　　　　　B. more phone lines to go over the same cable or more channels to come through

3. Digital signals　　　　　　　　　　C. transmit small-signal

4. Higher carrying capacity　　　　　　D. transmit in heat condition

6.9　Fiber to the Home

Fiber to the home (Fig. 6. 13) or FTTH is a telecommunications network based on optical fiber. Traditionally, most networks were based on a copper or coaxial cables. Optical fiber networks allow for a high volume of data transfer compared to other networks. This includes TV, Internet and phone services (so called tripple play).

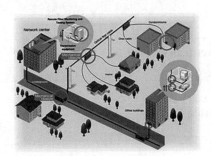

Fig. 6. 13　Fiber to the home

Over the last years, companies like Verizon and AT&T have deployed fiber networks across the US. Fiber to the home, or FTTH, means the fiber runs all the way to your home or apartment, vs e. g. FTTC or fiber to the curb where the fiber runs till your curb and the existing copper infrastructure is used for the last yards of data transfer (and hence forming the bottle neck).

Optical fiber has been used since a long time for long haul communications (e. g. the cables lying on the bottom of our ocean which take care of communications between the continents), however FTTH is a rather new development with high expectations. Many countries in Asia are leading in the access of FTTH by customers, leading regions are the Republic of Korea, Chinese Hong Kong and Japan. In Europe, especially the Nordic countries are leading. In the US, several companies are deploying a fiber to the home, FTTH, and have started with the most populated areas. The Obama government stimulus package contains incentives for the deployment of FTTH in rural areas. These areas are often served by small rural telecommunications companies. South Korea, in particular, is a world leader with more than 31% of its households boasting FTTH broadband connections. Chinese Hong Kong is second globally, with more than 23% penetration, while Japan is a close third with more than 21% of its households FTTH ready.

Although fiber-optic systems excel in high-bandwidth applications, optical fiber has been slow to achieve its goal of fiber to the home or to solve the last mile problem.

Notes

volume	*n.* 容量
haul	*n.* 距离
stimulus	*n.* 刺激
incentive	*n.* 激励
rural	*adj.* 农村的

Exercise

Directions: On the basis of the text, check out the link below:

http: //communication. howstuffworks. com/fiber-optic-communications/fiber-to-the-home3. htm

For more information about fiber-to-the-home broadband connections and related topics, then answer the questions, you can discuss with your partner.

1. What is tripple play?

_____

_____

2. Which country is the world leader in FTTH market penetration?

_____

_____

3. Which year is the best year in terms of new FTTH subscribers worldwide?

_____

_____

References

［1］ AGRAWAL G P. Fiber-optic communication systems ［M］. Hoboken：John Wiley&Sons，Inc.，2002.

［2］ OSCHE G R. Optical detection theory for laser applications ［M］. Hoboken：John Wiley&Sons，Inc.，2002.

［3］ STEEN W M，MAZUMDER J. Laser material processing ［M］. 4th ed. London：Springer，2010.

［4］ LAWRENCE J，Advances in laser materials processing：Technology，research and applications ［M］. Oxford：Woodhead Publishing Limited，2010.

［5］ LASER INSTITUTE OF AMERICA. Laser safety information bulletin ［EB］ Orlando：Laser Institute of America，2010.

［6］ TLF LASER. Laser processing ［A］ TLF Laser in Material Maching，2011.

［7］ ［美］KOECHNER W. 固体激光工程 ［M］. 孙文，江泽文，程国祥，译. 北京：科学出版社，2002.

［8］ ［美］约瑟夫・C・帕勒里斯. 光纤通信 ［M］. 北京：电子工业出版社，2005.

［9］ 王琳，夏怡. 电子与通信专业英语 ［M］ 北京：北京理工大学出版社，2007.

［10］ 刘小芹，刘骋. 电子与通信技术专业英语 ［M］. 2 版. 北京：人民邮电出版社，2008.

［11］ 李霞. 电子与通信专业英语 ［M］ 北京：电子工业出版社，2005.

［12］ 刘其斌. 激光加工技术及其应用 ［M］ 北京：冶金工业出版社，2007.

［13］ 教育部. 教育部关于全面提高高等职业教育教学质量的若干意见 ［EB/OL］. （2006 – 11 – 16）［2023 – 11 – 19］. http：//www. moe. gov. cn/srcsite/A07/s7055/200611/t20061116_79649. html.

［14］ 王九程，袁智英，黄丛笑. 高职《公共英语》的职业性及其提高探讨 ［J］. 湖北广播电视大学学报，2010，（11）：37 – 38.

［15］ 刘文宇，王慧莉，张旭. 商务英语口语大全 ［M］. 大连：大连理工大学出版社，2007.

［16］ 余啸海，薛昆. 实用英语口语速查手册 ［M］. 北京：国防工业出版社，2005.

［17］ 蒲村，刘华，燕静敏，等. 英语会话全程通 ［M］. 北京：机械工业出版社，2003.

［18］ 卢欣，王晓妍. 1 + X 脱口说：商务英语口语 ［M］. 大连：大连理工大学出版社，2007.

［19］ 张翠萍. 商贸英语口语大全 ［M］. 北京：对外经济贸易大学出版社，2006.

［20］ 盛小利. 商务英语谈判口语 ［M］. 北京：中国宇航出版社，2007.

［21］ 刘一平，李宏亮，商务英语口语 ［M］. 北京：北京大学出版社，2006.

［22］ 教育部《机电英语》教材编写组. 机电英语 ［M］. 北京：高等教育出版社，2001.

［23］ 赵萱，郑仰成. 科技英语翻译 ［M］. 北京；外语教学与研究出版社，2006.

［24］ 张朴顺. 现代服务业英语会话 900 句 ［M］. 北京：中国标准出版社，2005.

［25］ 谢江南，何加红. 实用英文写作 ［M］. 2 版. 北京：首都经济贸易大学出版社，2005.

［26］ 张道真. 实用英语语法 ［M］. 北京：外语教学与研究出版，1995.

［27］ 陈莉萍. 专门用途英语存在的依据 ［J］. 外语与外语教学，2001 （12）：28 – 30.

［28］ 范谊. ESP 存在的理据 ［J］. 外语教学与研究，1995 （3）：43 – 48.

［29］ 冯世梅，杜耀文. 现代科技英语词汇与翻译 ［J］，中国科技翻译，2002，15 （4）：52 – 54.

［30］ 冯志杰. 汉英科技翻译指要 ［M］. 北京：中国对外翻译出版公司，2001.

［31］ 郭海平. 科技英语词汇的构词特点及翻译 ［J］. 武汉轻工大学学报，2004，23 （2）：115. 117.

［32］ 刘双峰，王高. 光机电系统概论 ［M］. 北京北京理工大学出版社，2007.

［33］ 卫乃兴，周俊英. 也谈 ESP 与大学英语教学 ［J］. 外语界，1994 （2）：32 – 36.

［34］ 张少雄.“科技英语词汇”评说 ［J］. 外语教学与研究出版社，1994 （3）：47 – 50.

[35] 赵营，郑仰成. 科技英语翻译 [M]. 北京：外语教学与研究出版社，2006.

[36] 朱一纶. 电子技术专业英语 [M]. 北京：电子工业出版社，2006.

[37] 刘骋. 电子与通信技术专业英语 [M]. 北京：人民邮电出版社，2019.

Multimedia Resources Index

教材配套资源索引

序号	文件名称	页码	类别
1	Basic Circuit Concepts 思维导图	P27	思维导图
2	Electromagnetic Radiation 思维导图	P61	思维导图
3	Laser Construction 思维导图	P99	思维导图
4	Optical Detection Techniques 思维导图	P137	思维导图
5	Laser Marking 思维导图	P177	思维导图
6	Basic Circuit Concepts 微课	P27	微课
7	Electromagnetic Radiation 微课	P61	微课
8	Laser Construction 微课	P99	微课
9	Optical Detection Techniques 微课	P137	微课
10	Laser Marking 微课	P177	微课
11	Refraction 动画	P42	动画
12	Laser Construction 动画	P73	动画
13	Photoelectric Effect（car park model）动画	P113	动画
14	Optical Detection Techniques 动画	P121	动画
15	Laser Marking 动画	P154	动画

更多教材配套资源和练习登录智慧职教平台

学生：进入 https://www.icve.com.cn/的专业群，武汉职业技术学院的《光电技术应用专业英语》课程。

教师：进入 https://www.icve.com.cn/的职教云，搜索"光电技术应用专业英语"，组建课程。

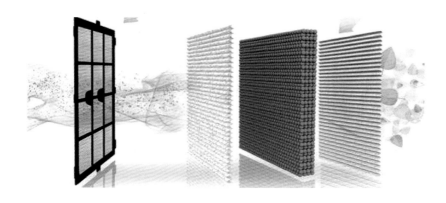

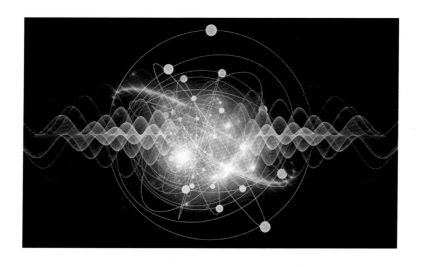

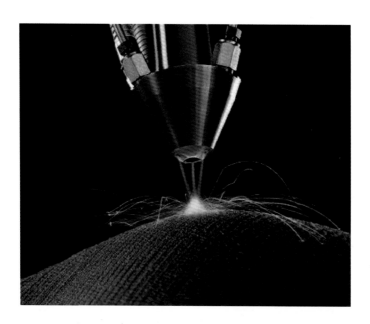

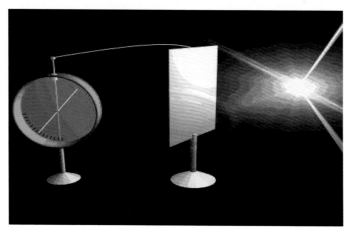

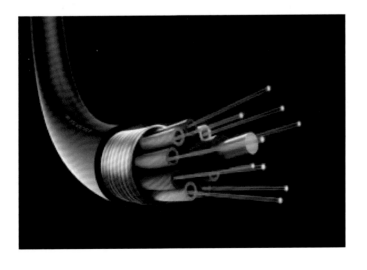

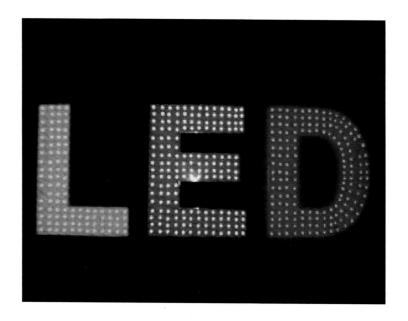